MODERN TOOLMAKING METHODS

A TREATISE ON PRECISION DIVIDING AND LOCATING METHODS, LAPPING, MAKING FORMING TOOLS, ACCURATE THREADING, BENCH LATHE PRACTICE, TOOLS FOR PRECISION MEASUREMENTS, AND GENERAL TOOLMAKING PRACTICE

COMPILED AND EDITED

BY

FRANKLIN D. JONES
ASSOCIATE EDITOR OF MACHINERY
AUTHOR OF "TURNING AND BORING"
AND "PLANING AND MILLING"

FIRST EDITION
SEVENTH PRINTING

Watchmaker Publishing
1915

PREFACE

As the work of the toolmaker requires an unusual degree of skill and refinement and is of great importance in connection with the modern system of interchangeable manufacture, it is surprising that so little has been published heretofore on toolmaking practice. This volume on "Modern Toolmaking Methods" is believed to meet a real need as it deals with a great variety of tool-room problems and explains many important toolmaking operations. Owing to the varied nature of toolmaking practice, no attempt has been made to cover completely every phase of work which might properly be classified as toolmaking. This treatise does, however, cover quite completely those methods and operations which are fundamental and essential to the production of small tools and precision work. It also contains many valuable mathematical rules and typical calculations that will aid in the solution of the practical problems which are so frequently encountered in the tool-room.

Some of the methods described represent standard practice, whereas others have been developed by different toolmakers for special operations, and may not prove to be the best under all conditions, because, as every mechanic knows, great accuracy is sometimes necessary regardless of the time required to do the work; whereas, in other cases, the time element is very important. Therefore, any one method may not always prove adequate and it is necessary for the toolmaker to consider the conditions in each case and be guided by his judgment and experience.

Readers of mechanical literature are familiar with MACHINERY's twenty-five cent Reference Books, of which one hundred and forty-one different titles have been published during the past seven years. As many subjects cannot be covered adequately in all their phases in books of this size, and in response to a demand for more comprehensive and detailed treatments of the more important mechanical subjects, it has been deemed advisable to publish a number of larger volumes, of which this is one. This treatise includes part of MACHINERY's Reference Books Nos. 31, 64, 130 and 135. The additional material was compiled largely from articles previously published in MACHIN-ERY. These were contributed by men engaged in many different branches of tool work so that the book is not a record of any one toolmaker's experience, but covers a broad field.

As accuracy is essential to practically all toolmaking operations, the various standard methods for locating, spacing and dividing precision work, have been made a special feature of this book. It also deals with such important subjects as lapping, production of straight and circular forming tools and formed cutters, fluting and grinding milling cutters and reamers, bench lathe practice, precision threading, gaging tools and methods, and many other subjects of practical value.

F. D. J.

NEW YORK, *February*, 1915.

CONTENTS

CHAPTER VII

PRECISION BENCH LATHE PRACTICE

CHAPTER VIII

GAGES AND MEASURING INSTRUMENTS

MODERN TOOLMAKING METHODS

CHAPTER I

PRECISION LOCATING AND DIVIDING METHODS

The degree of accuracy that is necessary in the construction of certain classes of machinery and tools, has made it necessary for toolmakers and machinists to employ various methods and appliances for locating holes or finished surfaces to given dimensions and within the prescribed limits of accuracy. In this chapter, various approved methods of locating work, such as are used more particularly in tool-rooms, are described and illustrated.

Button Method of Accurately Locating Work. — Among the different methods employed by toolmakers for accurately locating work such as jigs, etc., on the faceplate of a lathe, one of the most commonly used is known as the "button method." This method is so named because cylindrical bushings or buttons are attached to the work in positions corresponding to the holes to be bored, after which they are used in locating the work. These buttons which are ordinarily about $\frac{1}{2}$ or $\frac{5}{8}$ inch in diameter, are ground and lapped to the same size, and the ends are finished perfectly square. The outside diameter should preferably be such that the radius can easily be determined, and the hole through the center should be about $\frac{1}{8}$ inch larger than the retaining screw so that the button can be adjusted laterally.

As a simple example of the practical application of the button method, suppose three holes are to be bored in a jig-plate according to the dimensions given in Fig. 1. A common method of procedure would be as follows: First lay out the centers of all holes to be bored, by the usual method. Mark these centers with a prick-punch and then drill holes for the machine screws which are used to clamp the buttons. After the buttons are clamped lightly in place, set them in correct relation with each other and with the jig-plate. The proper location of the buttons is very important, as their positions largely determine the accuracy of the work. The best method of locating a number of buttons depends somewhat upon their relative positions, the instruments available, and the accuracy required. When buttons must be located at given distances from the finished sides of a jig, a surface plate and vernier height-gage are often used. The method is to place that side from which the button is to be

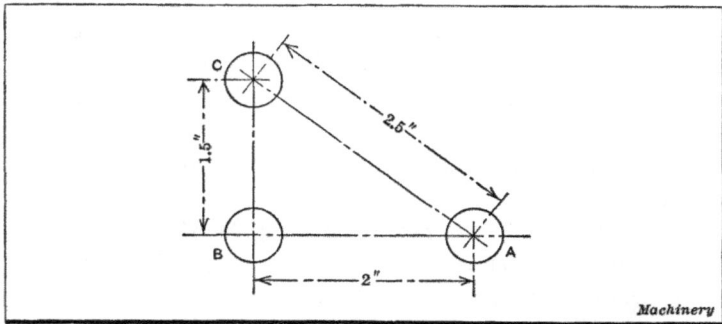

Fig. 1.   Simple Example of Work Illustrating Application of Button Method

set, upon an accurate surface plate and then set the button by means of the height-gage, allowance being made, of course, for the radius of the button. The center-to-center distance between the different buttons can afterwards be verified by taking direct measurements with a micrometer (as indicated in Fig. 2) by measuring the overall distance and deducting the diameter of one button.

After the buttons have been set and the screws are tightened, all measurements should be carefully checked. The work is then mounted on the faceplate of the lathe and one of the buttons is set true by the use of a test indicator. When the dial of the indicator ceases to vibrate, thus showing that the button runs true, the latter should be removed so that the hole can be drilled and bored to the required size. In a similar manner other buttons are indicated and the holes bored, one at a time. It is evident that if each button is correctly located and set perfectly true in the lathe, the various holes will be located the required distance apart within very close limits.

Another example of work illustrating the application of the button method is shown in Fig. 3. The disk-shaped part illus-

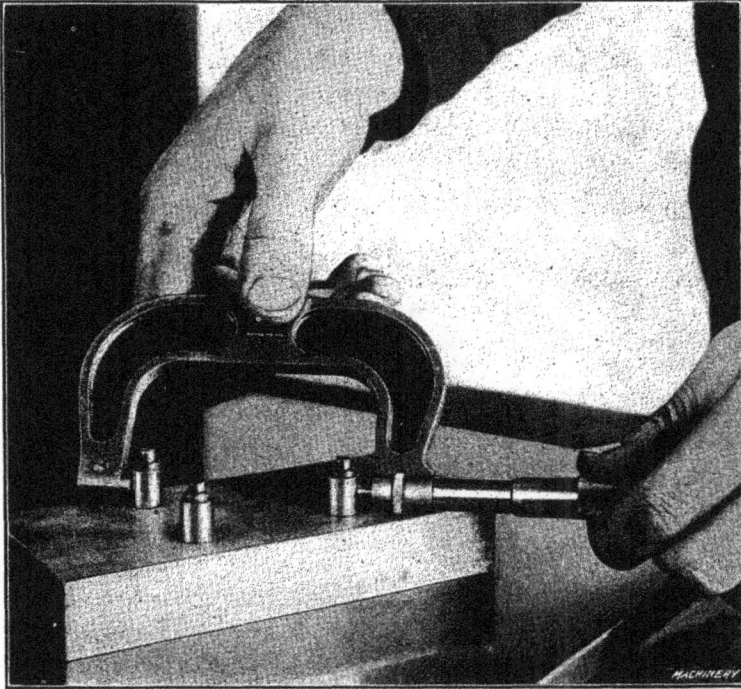

Fig. 2.   Testing Location of Buttons

trated is a flange templet which formed a part of a fixture for drilling holes in flanged plates, the holes being located on a circle 6 inches in diameter. It was necessary to space the six holes equi-distantly so that the holes in the flanges would match in any position, thus making them interchangeable. First a plug was turned so that it fitted snugly in the $1\frac{1}{4}$-inch central hole of the plate and projected above the top surface about $\frac{3}{4}$ inch. A center was located in this plug and from it a circle

of three inches radius was drawn.   This circle was divided into
six equal parts and then small circles $\frac{5}{8}$ inch in diameter were
drawn to indicate the outside circumference of the bushings to
be placed in the holes.   These circles served as a guide when
setting the button and enabled the work to be done much more
quickly.   The centers of the holes were next carefully prick-
punched and small holes were drilled and tapped for No. 10
machine screws.   After this the six buttons were attached in
approximately the correct positions and the screws tightened
enough to hold the buttons firmly, but allow them to be moved
by tapping lightly.   As the radius of the circle is 3 inches, the

Fig. 3.   Flange Templet with Buttons Attached

radius of the central plug, $\frac{5}{8}$ inch, and that of each button, $\frac{5}{16}$
inch, the distance from the outside of the central plug to the
outside of any button, when correctly set, must be $3\frac{5}{16}$ inches.
Since there are six buttons around the circle, the center-to-center
distance is equal to the radius, and the distance between the
outside or any two buttons should be $3\frac{5}{8}$ inches.   Having determined these dimensions, each button is set equi-distant from
the central plug and the required distance apart, by using a
micrometer.   As each button is brought into its correct position,
it should be tightened down a little so that it will be located
firmly when finally set.   The work is then strapped to the faceplate of a lathe and each button is indicated for boring the
different holes by means of an indicator, as previously described.

When the buttons are removed it will be found that in nearly all cases the small screw holes will not run exactly true; therefore it is advisable to form a true starting point for the drill by using a lathe tool.

When doing precision work of this kind, the degree of accuracy obtained will depend upon the instruments used, the judgment and skill of the workman, and the care exercised. A good general rule to follow when locating bushings or buttons is to use the method which is the most direct and which requires the least number of measurements, in order to prevent an accumulation of errors.

Locating Work by the Disk Method. — Comparatively small precision work is sometimes located by the disk method, which is the same in principle as the button method, the chief difference being that disks are used instead of buttons. These disks are made to such diameters that when their peripheries are in contact, each disk center will coincide with the position of the hole to be bored; the centers are then used for locating the work. To illustrate this method, suppose that the masterplate shown at the left in Fig. 4 is to have three holes a, b and c bored into it, to the center distances given.

It is first necessary to determine the diameters of the disks. If the center distances between all the holes were equal, the diameters would, of course, equal this dimension. When, however, the distances between the centers are unequal, the diameters may be found as follows: Subtract, say, dimension y from x, thus obtaining the difference between the radii of disks C and A (see right-hand sketch); add this difference to dimension z, and the result will be the diameter of disk A. Dividing this diameter by 2 gives the radius, which, subtracted from center distance x equals the radius of B; similarly, the radius of B subtracted from dimension y equals the radius of C.

For example, $0.930 - 0.720 = 0.210$ or the difference between the radii of disks C and A. Then the diameter of $A = 0.210 + 0.860 = 1.070$ inch, and the radius equals $1.070 \div 2 = 0.535$ inch. The radius of $B = 0.930 - 0.535 = 0.395$ inch and $0.395 \times 2 = 0.790$, or the diameter of B. The center distance

0.720 − 0.395 = 0.325, which is the radius of C; 0.325 × 2 = 0.650 or the diameter of C.

After determining the diameters, the disks should be turned nearly to size and finished, preferably in a bench lathe. First insert a solder chuck in the spindle, face it perfectly true, and attach the disk by a few drops of solder, being careful to hold the work firmly against the chuck while soldering. Face the outer side and cut a sharp V-center in it; then grind the periphery to the required diameter. Next fasten the finished disks onto the work in their correct locations with their peripheries in contact, and then set one of the disks exactly central with the lathe

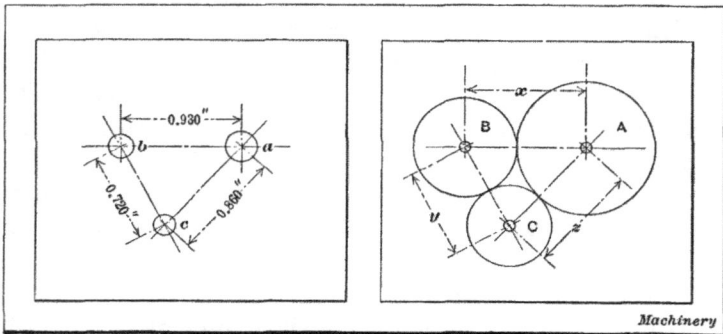

Fig. 4. An Example of Precision Work, and Method of Locating Holes by Use of Disks in Contact

spindle by applying a test indicator to the center in the disk. After removing the disk and boring the hole, the work is located for boring the other holes in the same manner.

Small disks may be secured to the work by means of jeweler's wax. This is composed of common rosin and plaster of Paris and is made as follows: Heat the rosin in a vessel until it flows freely, and then add plaster of Paris and keep stirring the mixture. Care should be taken not to make the mixture too stiff. When it appears to have the proper consistency, pour some of it onto a slate or marble slab and allow it to cool; then insert the point of a knife under the flattened cake thus formed and try to pry it off. If it springs off with a slight metallic ring, the proportions are right, but if it is gummy and ductile, there is too much rosin. On the other hand, if it is too brittle and crumbles, this indicates that there is too much plaster of Paris. The wax should be warmed before using. A mixture of beeswax and shellac, or beeswax and rosin in about equal proportions, is also used for holding disks in place. When the latter are fairly large, it may be advisable to secure them with small screws, provided the screw holes are not objectionable.

Disk-and-button Method of Locating Holes. — The accuracy of work done by the button method previously described is

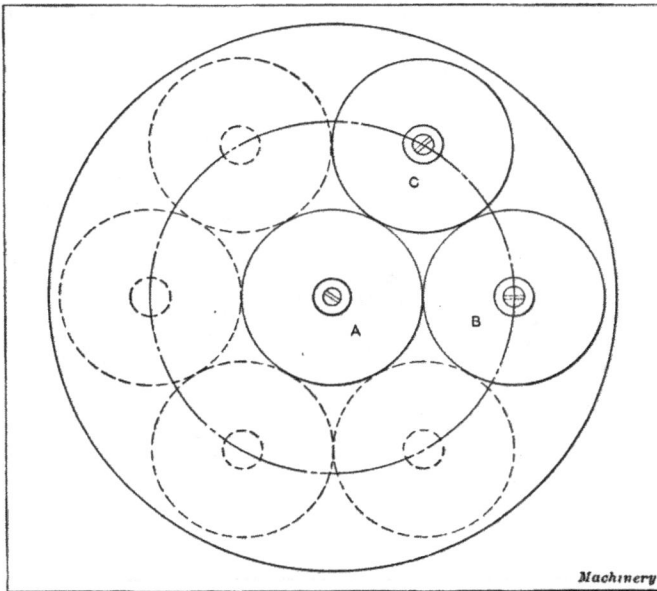

Fig. 5. Locating Equi-distant Holes on a Circle by Using Disks and Buttons in Combination

limited only by the skill and painstaking care of the workman, but setting the buttons requires a great deal of time. By a little modification, using what is sometimes called the "disk-and-button method," a large part of this time can be saved without any sacrifice of accuracy. The disk-and-button method, which was described by Guy H. Gardner in MACHINERY, September, 1914, is extensively used in many shops. Buttons are used, but they are located in the centers of disks of whatever diameters are necessary to give the required locations. As three disks are used in each step of the process, it is sometimes called the "three-disk method."

To illustrate the practical application of this method, suppose six equally-spaced holes are to be located in the circumference of a circle six inches in diameter. To locate these, one needs, besides the buttons, three disks three inches in diameter, each having a central hole exactly fitting the buttons. It is best to have, also, a bushing of the same diameter as the buttons, which has a center-punch fitted to slide in it. First, the center button

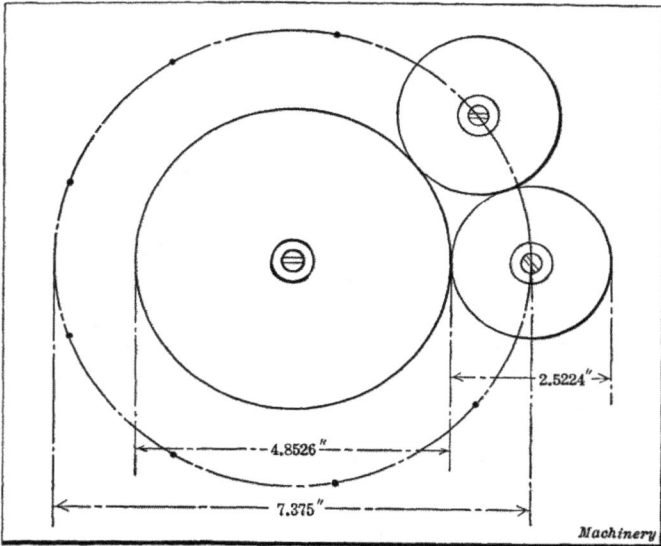

Fig. 6. **Example of Circular Spacing Requiring a Large Central Disk**

is screwed to the templet, and one of the disks *A*, Fig. 5, is slipped over it; then a second disk *B* carrying a bushing and center-punch is placed in contact with disk *A* and a light blow on the punch marks the place to drill and tap for No. 2 button, which is kept in its proper place while tightening the screw by holding the two disks *A* and *B* in contact. Next, the third disk *C* is placed in contact with disks *A* and *B* and locates No. 3 button, and so on until the seven buttons are secured in position. The templet is then ready to be strapped to the lathe faceplate for boring.

Of course, it is not possible to use disks of "standard" sizes for many operations, but making a special disk is easy, and its cost is insignificant as compared with the time saved by its use. One who employs this method, especially if he also uses disks to lay out angles, soon accumulates a stock of various sizes. While it is desirable to have disks of tool steel, hardened and ground, or, in the larger sizes, of machine steel, casehardened and ground, a disk for occasional use will be entirely satisfactory if left soft.

Two other jobs that illustrate this method may be of interest. The first one, shown in Fig. 6, required the locating of nine equally-spaced holes on a circumference of $7\frac{3}{8}$-inches diameter. In any such case, the size of the smaller disks is found by multiplying the diameter of the circle upon which the centers of the disks are located by the sine of half the angle between two adjacent disks. The angle between the centers of adjacent disks equals 360 ÷ number of disks. 360 ÷ 9 = 40; hence, in this case, the diameter of the smaller disks equals $7\frac{3}{8}$ multiplied by the sine of 20 degrees, or $7\frac{3}{8} \times 0.34202 = 2.5224$ inches. $7\frac{3}{8} - 2.5224 = 4.8526$ inches, which is the diameter of the central disk.

The templet shown in Fig. 7 required two holes on a circumference $6\frac{1}{2}$-inches diameter, with their centers 37 degrees 20 minutes apart. To find the diameter of the smaller disks, multiply the diameter of the large circle by the sine of one-half the required angle, as in the preceding example; thus $6\frac{1}{2} \times \sin$ 18 degrees 40 minutes = 2.0804 inches, which is the diameter of the two smaller disks. The diameter of the larger disk equals $6\frac{1}{2} - 2.0804 = 4.4196$ inches.

Very accurate results can be obtained by the disk-and-button method. Of course, absolute exactness is equally unattainable with buttons and a micrometer, or any other method; the micrometer does not show the slight inaccuracy in any one chordal measurement, while in using the disks the error is accumulative and the insertion of the last disk in the series shows the sum of the errors in all the disks. It is only in cases like the one illustrated in Fig. 5 that we note this, and then, though in correcting the error we may change the diameter of the circle a very slight amount, an exceedingly accurate division of the circumference is secured.

Use of Two- and Three-diameter Disks. — Fig. 8 illustrates, on an enlarged scale, a piece of work requiring great accuracy, which was successfully handled by an extension of the three-disk method.   Fourteen holes were required in a space hardly larger than a silver half-dollar, and, although the drawing gave dimensions from the center of the circle, the actual center could not be used in doing the work, as there was to be no hole there; moreover, a boss slightly off center prevented the use of a central disk, unless the bottom of the disk were bored out to receive this boss, which was not · thought expedient.   Hence, the method adopted was to make the plate thicker than the dimension given on the drawing, and then bore it out to leave a rim of definite diameter, this rim to be removed after it had served its purpose as a locating limit for the disks.

As the holes A and B, which were finished first, were 0.600 inch apart and 0.625 inch from the center, the rim was bored to 1.850 inch and two 0.600-inch disks, in contact with the rim and with each other, located these holes.   As hole C was to be equidistant from holes A and B, and its distance from the center was given, the size of the disk for this hole was readily determined. The disks for holes A, B and C have two diameters; the upper diameters are made to whatever size is required for locating the disks of adjacent holes, and they also form a hub which can be used when setting the disks with an indicator.   Hole D was 0.4219 inch from B, and calculations based on this dimension and its distance from the center showed that it was 0.4375 inch from hole C.

A "three-story" disk or button was made for hole D.   The diameter of the large part was 0.46875 inch and it overlapped· disks C and B (the upper sections of which were made 0.375 inch and 0.4062 inch, respectively), thus locating D.   Then hole F and all the remaining holes were located in a similar manner. The upper diameters of disks E and D were used in locating disks for other adjacent holes, as well as a hub for the indicator; for

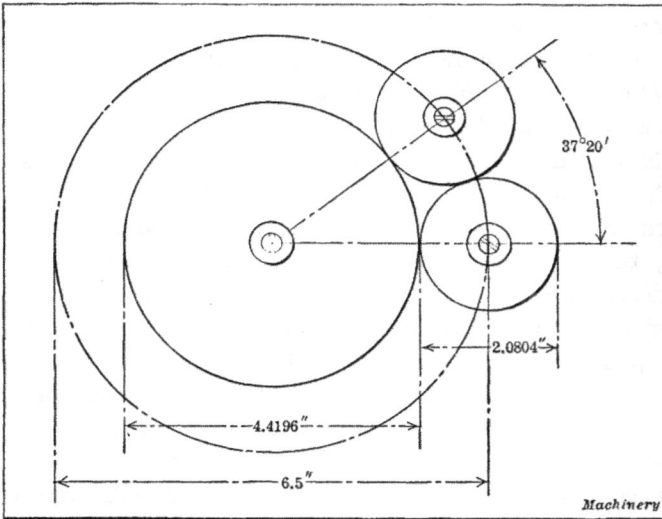

Fig. 7. Locating Holes at an Angle by Use of Disks and Buttons

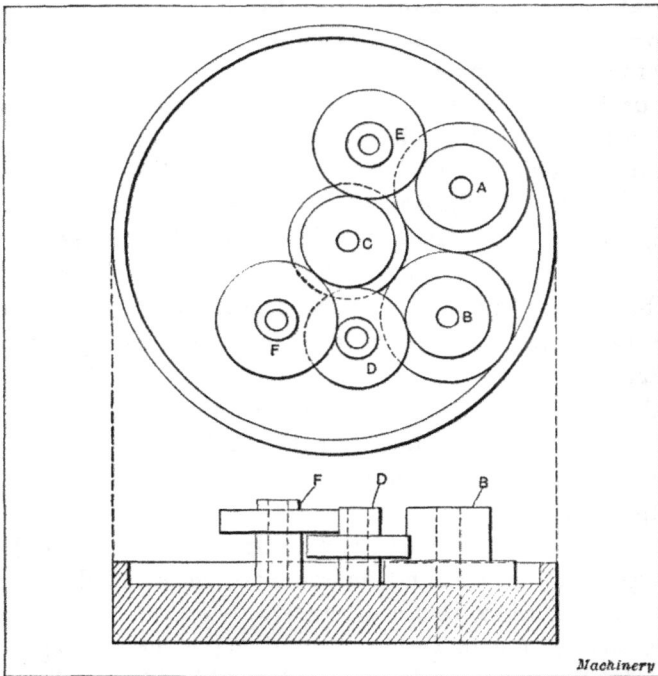

Fig. 8. Locating Holes by Means of Two- and Three-diameter Disks
in Contact

instance, to locate a hole with reference to holes C and D, the diameter of the new disk and the diameter of the upper part of disk D were varied to give the required location. The relation between the disks B, D and F is shown by the side view.

It had been decided that no screws should be used in attaching the buttons or disks to the work, as it was feared that the tapped holes would introduce inaccuracy by deflecting the boring tools; therefore, the following method was employed. After all the disks were fastened in place by clamps, a soft solder of low melting point was flowed about them, filling the work to the top of the rim. When the solder had cooled, the clamps were removed, the work transferred to the lathe faceplate, indicated in the usual way, and the holes bored by a "D" or "hog-nose" drill, guided by an axial hole in each disk, which had been provided for that purpose when the disks were made. It was thought that the unequal contraction of the solder and the plate in cooling might throw the holes "out of square;" however, careful measurements failed to show any appreciable lack of parallelism in test-bars inserted in the holes.

Accurate Angular Measurements with Disks. — For setting up a piece of work on which a surface is to be planed or milled at an exact angle to a surface already finished, disks provide an accurate means of adjustment. One method of using disks for angular work is illustrated at A in Fig. 9. Let us assume that the lower edge of plate shown is finished and that the upper edge is to be milled at an angle α of 32 degrees with the lower edge. If the two disks x and y are to be used for locating the work, how far apart must they be set in order to locate it at the required angle? The center-to-center distance can be determined as follows: Subtract the radius of the smaller disk from the radius of the larger disk, and divide the difference by the sine of one-half the required angle.

Example. — If the required angle α is 32 degrees, the radius of the large disk 2 inches, and the radius of the small disk 1 inch, what is the center-to-center distance?

The sine of one-half the required angle, or 16 degrees, is 0.27564. The difference between the radii of the disks equals

2 — 1 = 1, and 1 ÷ 0.27564 = 3.624 inches. Therefore, for an angle of 32 degrees, disks of the sizes given should be set so that the distance between their centers is 3.624 inches.

Another method of accurately locating angular work is illustrated at *B* in Fig. 9. In this case, two disks are also used, but they are placed in contact with each other and changes for different angles are obtained by varying the diameter of the larger disk. The smaller disk is a standard 1-inch size, such as is used for setting a 2-inch micrometer. By this method any angle up to about 40 degrees can be obtained within a very close limit of accuracy. The following rule may be used for determin-

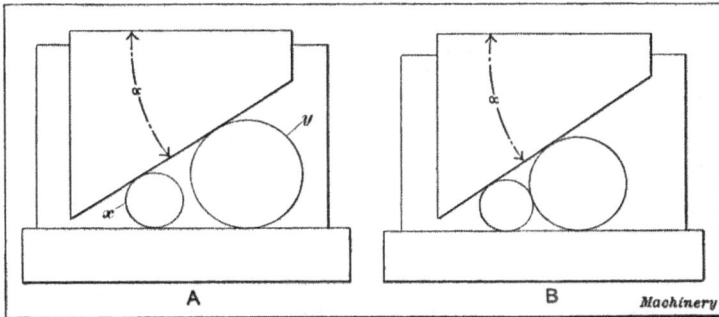

Fig. 9. **Obtaining Accurate Angular Measurements with Disks**

ing the diameter of the larger disk, when both disks are in contact and the diameter of the small disk is known:

Multiply twice the diameter of the small disk by the sine of one-half the required angle; divide this product by 1 minus the sine of one-half the required angle; add the quotient to the diameter of the small disk to obtain the diameter of the large disk.

Example. — The required angle α is 15 degrees. Find the diameter of the large disk to be in contact with the standard 1-inch reference disk.

The sine of 7 degrees 30 minutes is 0.13053. Multiplying twice the diameter of the small disk by the sine of 7 degrees 30 minutes, we have 2 × 1 × 0.13053 = 0.26106. This product divided by 1 minus the sine of 7 degrees 30 minutes = $\dfrac{0.26106}{1 - 0.13053}$ = 0.3002.

This quotient added to the diameter of the small disk equals
1 + 0.3002 = 1.3002 inch, which is the diameter of the large disk.

The accompanying table gives the sizes of the larger disks to
the nearest 0.0001 inch for whole degrees ranging from 5 to 40

Disk Diameters for Angular Measurement

Deg.	Inch.	Deg.	Inch.	Deg.	Inch.	Deg.	Inch.
5	1.0912	14	1.2775	23	1.4980	32	1.7610
6	1.1104	15	1.3002	24	1.5249	33	1.7934
7	1.1300	16	1.3234	25	1.5524	34	1.8262
8	1.1499	17	1.3468	26	1.5805	35	1.8600
9	1.1702	18	1.3708	27	1.6090	36	1.8944
10	1.1909	19	1.3953	28	1.6382	37	1.9295
11	1.2120	20	1.4203	29	1.6680	38	1.9654
12	1.2334	21	1.4457	30	1.6983	39	2.0021
13	1.2553	22	1.4716	31	1.7294	40	2.0396

degrees inclusive. Incidentally, the usefulness of these disks
can be increased by stamping on each one its diameter and also
the angle which it subtends when placed in contact with the
standard 1-inch disk.

Disk-and-square Method of Determining Angles. — The
method shown in Fig. 10 for determining angles for setting up
work on a milling machine or planer, possesses several advantages. No expensive tools are required, the method can be applied quickly, and the results obtained are quite accurate enough
for any but the most exacting requirements. As will be seen,
an ordinary combination square is used in connection with a
disk, the head of the square being set at different points on the
blade according to the angle that is desired. Theoretically, a
one-inch disk could be used for all angles from about 6 degrees
up to a right angle, but in practice it is more convenient and
accurate to employ larger disks for the larger angles.

The only inaccuracy resulting from this method is due to
setting the square at the nearest "scale fraction" instead of at
the exact point determined by calculation. This error is very
small, however, and is negligible in practically all cases. The
dimension x required for any desired angle α can be found by
multiplying the radius of the disk by the cotangent of one-half

the desired angle, and adding to this product the radius of the disk.

Example. — The square blade is to be set to an angle of 15 degrees 10 minutes, using a 2-inch disk. At what distance x (see Fig. 10) should the head of the square be set?

Cot 7 degrees 35 minutes = 7.5113, and 7.5113 × 1 + 1 = 8.5113 inches.

By setting the square to $8\frac{1}{2}$ inches "full," the blade would be set very close to the required angle of 15 degrees 10 minutes.

Locating Work by Means of Size Blocks. — The size-block method of locating a jig-plate or other part, in different positions on a lathe faceplate, for boring holes accurately at given center-

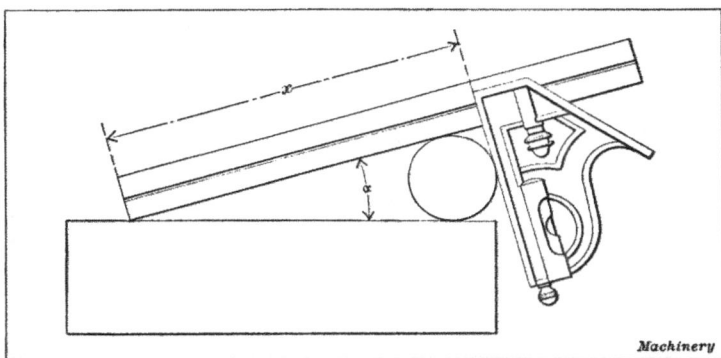

Fig. 10.   **Disk-and-square Method of Accurately Setting Angular Work**

to-center distances, is illustrated in Fig. 11.   The way the size blocks are used in this particular instance is as follows: A pair of accurate parallels are attached to a faceplate at right angles to each other and they are so located that the center of one of the holes to be bored will coincide with the lathe spindle.   The hole which is aligned in this way should be that one on the work which is nearest the outer corner, so that the remaining holes can be set in a central position by adjusting the work away from the parallels.   After the first hole is bored, the work is located for boring each additional hole by placing size blocks of the required width between the edges of the work and the parallels. For instance, to set the plate for boring hole D, size blocks (or a combination of blocks or gages) equal in width to dimension A_1

would be inserted at A, and other blocks equal in width to dimension B_1 beneath the work as at B. As will be seen, the dimensions of these blocks equal the horizontal and vertical distances between holes C and D. With the use of other combinations of gage blocks, any additional holes that might be required are located in the central position. While only two holes are shown in this case, it will be understood that the plate could be located accurately for boring almost any number of holes by this method.

Incidentally, such gages as the Johansson combination gages are particularly suited for work of this kind, as any dimension within the minimum and maximum limits of a set can be obtained

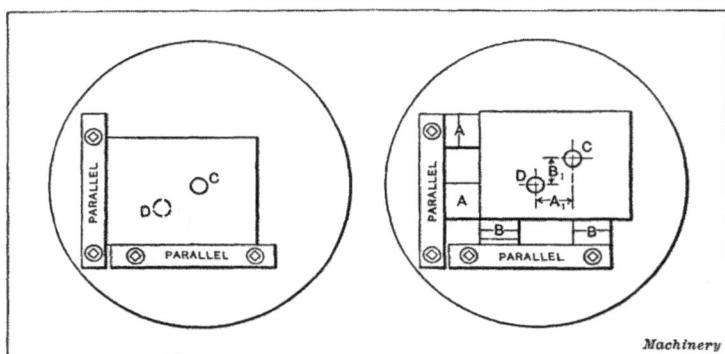

Fig. 11.   Method of Setting Work on Faceplate with Size Blocks or Gages

by simply placing the required sizes together. Sometimes, when such gages are not available, disks which have been ground to the required diameter are interposed between the parallels and the work for securing accurate locations. Another method of securing a positive adjustment of the work is to use parallels composed of two tapering sections, which can be adjusted to vary the width and be locked together by means of screws. Each half has the same taper so that outer edges are parallel for any position, and the width is measured by using a micrometer.

The size-block method is usually applied to work having accurately machined edges, although a part having edges which are of a rough or irregular shape can be located by this method, if it is mounted on an auxiliary plate having accurately finished square edges.   For instance, if holes were to be bored in the casting for a jig templet which had simply been planed on the top and bottom, the casting could be bolted to a finished plate having square edges and the latter be set in the different positions required by means of size blocks.   Comparatively large jig-plates are sometimes located for boring in this way and the milling machine is often used instead of a lathe.

The Master-plate Method. — When it is necessary to machine two or more plates so that they are duplicates as to the location of holes, circular recesses, etc., what is known as a master-plate is often used for locating the work on the lathe faceplate.   This master-plate M (see Fig. 12) contains holes which correspond to

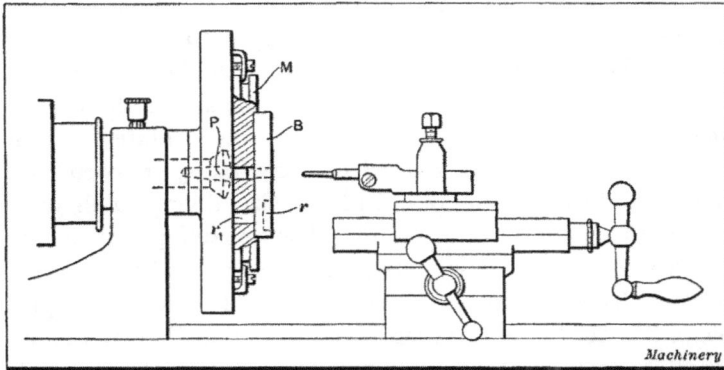

Fig. 12.   Master-plate applied to a Bench Lathe Faceplate

those wanted in the work, and which accurately fit a central plug P in the lathe spindle, so that by engaging first one hole and then another with the plug, the work is accurately positioned for the various operations.

When making the master-plate, great care should be taken to have the sides parallel and the holes at right angles to the sides, as well as accurately located with reference to one another.   The various holes may be located with considerable precision by the use of buttons as previously described.   Of course, it is necessary to have a hole in the master-plate for each different position in which the work will have to be placed on the faceplate; for example, if a circular recess r were required, a hole r_1 exactly concentric with it would be needed in the master-plate.   The method of holding the work and locating it with reference to the holes in the master-plate will depend largely on its shape.   The cylindrical blank B illustrated is positioned by a recess in the master-plate in which it fits.   The work is commonly held to the master-plate by means of clamps and tap bolts or by screws which pass through the work and into the master-plate.   Solder is sometimes used when it is not convenient to hold the work by clamps or screws.

The plug P which locates the master-plate is first turned to fit the spindle or collet of the lathe, and the outer or projecting end is rough-turned for the holes in the master-plate, which should all be finished to exactly the same diameter.   The plug is then inserted in the spindle and ground and lapped to a close fit for the holes in the master-plate.   The latter, with the work attached to it, is next clamped to the faceplate by the straps shown, which engage a groove around the edge of the master-plate.   The first hole is finished by drilling to within, say, 0.005 or 0.006 inch of the size, and then boring practically to size, a very small amount being left for reaming or grinding.   The remaining holes can then be finished in the same way, the work being positively located in each case by loosening the master-plate and engaging the proper hole in it with the central plug.   It is apparent that by the use of this same master-plate, a number of pieces B could be made which would be practically duplicates.

The master-plate method of locating work can be applied in many different ways.   It is used for making duplicate dies, for accurately locating the various holes in watch movements, and for many other operations requiring great precision.   Master-plates are quite frequently used by toolmakers when it is necessary to produce a number of drill jigs which are to be used for drilling holes in different parts having the same relative locations, thus requiring jigs that are duplicates within very close limits.

Making Master-plate for Duplicating a Model. — When a master-plate is required, that is to be used in making duplicates of an existing model, the holes are bored in the master-plate by reversing the process illustrated in Fig. 12.   That is, the central plug *P* is turned to fit the largest hole in the model and the latter with the attached master-plate blank is clamped to lathe face-plate. The first hole is then bored to within say 0.002 inch of the finish diameter, to allow for grinding, provided the master-plate is to be hardened. The central plug is then turned down to fit the next largest hole and the second hole is bored in the master-plate. This method is continued until all the holes are bored. In order to prevent any change in the position of the master-plate relative to the model, it may be secured by inserting dowel-pins through both parts, the work being held to the lathe faceplate by ordinary screw clamps. If the holes in the model do not extend clear through, a flat plate having parallel sides may

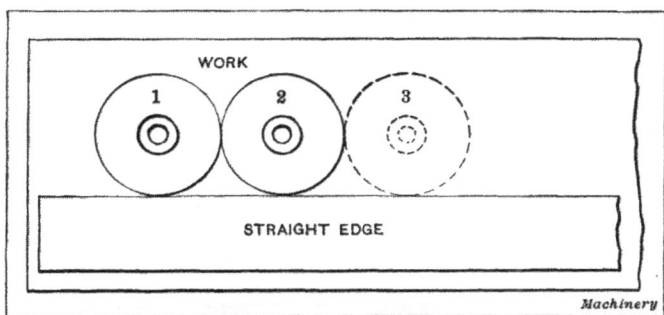

Fig. 13. Locating Equi-distant Holes in a Straight Line by Means of Disks and Straightedge

be interposed between the model and master-plate to provide clearance between the two and prevent cutting into the model when boring the master-plate.

Use of Disks for Locating Equally-spaced Holes. — A simple method of spacing holes that are to be drilled in a straight line is illustrated in Fig. 13. Two disks are made, each having a diameter equal to the center-to-center distance required between the holes. These disks must also have holes which are exactly central with the outside to act as a guide for the drill or reamer. The first two holes are drilled in the work while the disks are clamped so that they are in contact with each other and also with the straightedge as shown. One disk is then placed on the opposite side of the other, as indicated by the dotted line, and a third hole is drilled; this process of setting one disk against the opposite side of the other is continued until all the holes are drilled. When it is necessary to drill a parallel row of "staggered" holes, the second row can be located by placing disks of the proper size in contact with the first row of disks.

A method of using disks, which is preferable for very accurate work, is shown in Fig. 14. The disks are clamped against each other and along straightedge A by the screws shown, and if the outside diameters are correct and the guide holes concentric with the outside, very accurate work can be done. With this device there may be as many disks as there are holes to be drilled, if the number of holes is comparatively small, but if it is necessary to drill a long row of holes, the disks and frame are shifted along an

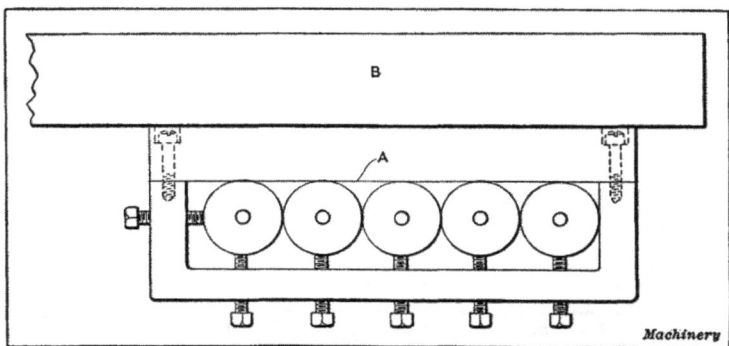

Fig. 14.   Special Disk-jig for Precision Drilling

auxiliary straightedge B, the hole in one of the end disks being aligned with the last hole drilled by inserting a close-fitting plug through the disk and hole.

Adjustable Jig for Accurate Hole Spacing. — An adjustable jig for accurately spacing small holes is shown in Fig. 15. This form is especially adapted for locating a number of equally-spaced holes between two previously drilled or bored holes, and the accuracy of the method lies in the fact that a slight error in the original spacing of the guide bushing is multiplied, and, therefore, easily detected. There are two of these guide bushings A and B which are carried by independent slides. These slides can be shifted along a dovetail groove after loosening the screws of clamp-gib C. To illustrate the method of using this jig, suppose five equally-spaced holes are to be located between two holes that are 12 inches apart. As the center-to-center distance between adjacent holes is 2 inches, slides A and B would be set so that the dimension x equals 2 inches plus the radii of the bushings. A straightedge is then clamped to the work in such position that a close-fitting plug can be inserted through the end holes which were previously drilled or bored. Then with a plug inserted through, say, bushing B and one of the end holes, the first hole is drilled and reamed through bushing A; the jig is then shifted to the left until the plug in B enters the hole just made. The second hole is then drilled and reamed through bushing A and this drilling and shifting of the jig is continued until the last hole

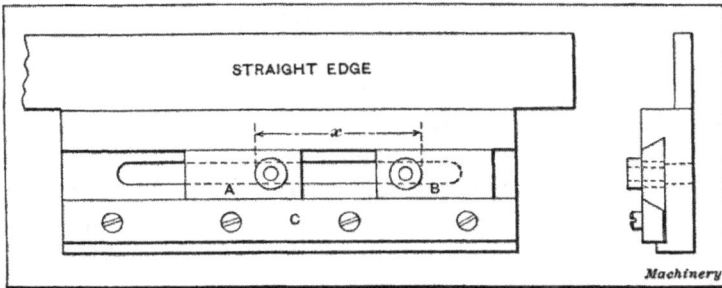

Fig. 15. Adjustable Jig for Accurate Hole Spacing

is finished. The distance between the last hole and the original end hole at the left is next tested by attempting to insert close-fitting plugs through both bushings. Evidently, if there were any inaccuracy in the spacing of the bushings, this would be multiplied as many times as the jig was shifted, the error being accumulative. To illustrate how the error accumulates, suppose that the bushings were 0.001 inch too far apart; then the distance to the first hole would be 2.001 inch, to the second hole, 4.002 inch, and finally the distance from the first to the sixth hole would be 10.005 inches; consequently, the distance between the sixth and seventh holes would equal 12 − 10.005 = 1.995 inch, or 0.005 inch less than the required spacing, assuming, for the sake of illustration, that the first and last holes were exactly 12 inches apart. In case of an error of 0.005 inch, the bushings would be set closer together an amount equal to one-fifth of this error, as near as could be determined with a micrometer, and all of the holes would then be re-reamed.

Methods of Accurately Dividing a Circle. — Sometimes it is necessary to machine a number of holes in a plate so that all the holes are on a circle or equi-distant from a central point, and also the same distance apart, within very small limits. A simple method of spacing holes equally is illustrated at *A*, Fig. 16. A number of buttons equal to the number of holes required are ground and lapped to exactly the same diameter, preferably by mounting them all on an arbor and finishing them at the same time. The ends should also be made square with the cylindrical surface of the button. When these buttons are finished, the diameter is carefully measured and this dimension is subtracted from the diameter of the circle on which the holes are to be located in order to obtain the diameter *d* (see illustration). A narrow shoulder is then turned on the plate to be bored, the diameter being made exactly equal to dimension *d*. By placing the buttons in contact with this shoulder, they are accurately located radially and can then be set equi-distant from each other by the use of a micrometer. In this particular case, it would be advisable to begin by setting the four buttons which are 90 degrees apart and then the remaining four. The buttons are next used for setting the work preparatory to boring.

Correcting Spacing Errors by Split Ring Method. — Another method of securing equal spacing for holes in indexing wheels, etc., is illustrated at *B*, Fig. 16. This method, however, is not to be recommended if the diameter of the circle on which the holes are to be located must be very accurate. The disk or ring in which the holes are required is formed of two sections *e* and *f*, instead of being one solid piece. The centers for the holes are first laid out as accurately as possible on ring *e*. Parts *e* and *f* are then clamped together and the holes are drilled through these two sections. Obviously, when the holes are laid out and drilled in this way, there will be some error in the spacing, and, consequently, all of the holes would not match, except when plate *e* is in the position it occupied when being drilled. Whatever errors

may exist in the spacing can be eliminated, however, by successively shifting plate *e* to different positions and re-reaming the
holes for each position. A taper reamer is used and two pins
should be provided having the same taper as the reamer. Ring
e is first located so that a hole is aligned quite accurately with one
in the lower plate. The ring is then clamped and the hole is
partly reamed, the reamer being inserted far enough to finish the
hole in plate *e* and also cut clear around in the upper part of plate

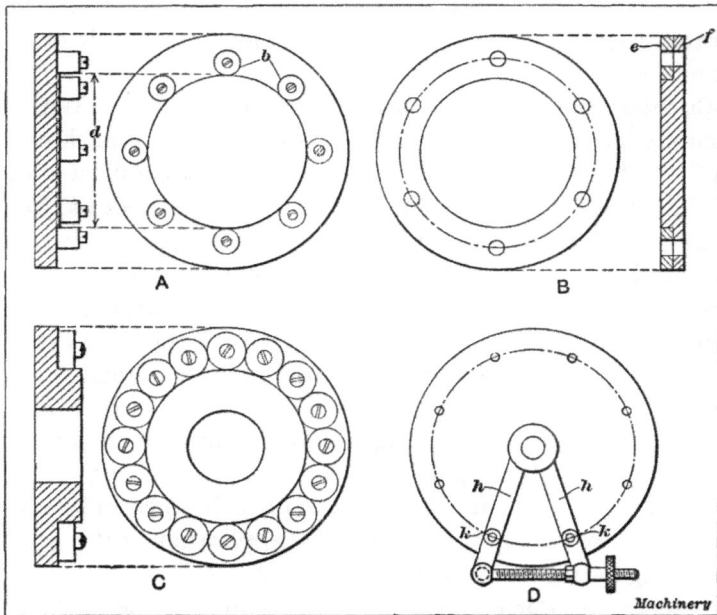

Fig. 16. Four Methods of Accurately Dividing a Circle

f. One of the taper pins is then driven into this hole and then a
hole on the opposite side is partly reamed, after which the other
pin is inserted. The remaining holes are now reamed in the same
way, and the reamer should be fed in to the same depth in each
case. If a pair of holes is considerably out of alignment, it may
be necessary to run the reamer in to a greater depth than was required for the first pair reamed, and in such a case all the holes
should be re-reamed to secure a uniform size.

The next step in this operation is to remove the taper pins and

clamps, turn index plate e one hole and again clamp it in position. The reaming process just described is then repeated; the
holes on opposite sides of the plate are re-reamed somewhat
deeper, the taper pins are inserted, and then all of the remaining
holes are re-reamed to secure perfect alignment for the new position of the plate. By repeating this process of shifting plate e
and re-reaming the holes, whatever error that may have existed
originally in the spacing of the holes, will practically be eliminated. It would be very difficult, however, to have these holes
located with any great degree of accuracy, on a circle of given
diameter.

Circular Spacing by Contact of Uniform Disks. — When an
accurate indexing or dividing wheel is required on a machine, the
method of securing accurate divisions of the circle illustrated at
C, Fig. 16, is sometimes employed. There is a series of circular
disks or bushings equal in number to the divisions required, and
these disks are all in contact with each other and with a circular
boss or shoulder on the plate to which they are attached. The
space between adjacent disks is used to accurately locate the
dividing wheel, engagement being made with a suitable latch or
indexing device. When making a dividing wheel of this kind, all
of the disks are ground and lapped to the same diameter and then
the diameter of the central boss or plate is gradually reduced
until all of the disks are in contact with each other and with the
boss. For an example of the practical application of this method
see "Originating a Precision Dividing Wheel."

Spacing by Correcting the Accumulated Error. — Another
indexing method of spacing holes equi-distantly, is illustrated
by the diagram at D, Fig. 16. An accurately fitting plug is inserted in the central hole of the plate in which holes are required.
Two arms h are closely fitted to this plug, but are free to rotate
and are provided with a fine-pitch screw and nut at the outer
ends for adjusting the distance between the arms. Each arm
contains an accurately-made, hardened steel bushing k located
at the same radial distance from the center of the plate. These
bushings are used as a guide for the drill and reamer when machining the holes in the plate.

To determine the center-to-center distance between the bushings, divide 360 by twice the number of holes required; find the sine corresponding to the angle thus obtained, and multiply it by the diameter of the circle upon which the holes are located. For example, if there were to be eleven holes on a circle 8 inches in diameter, the distance between the centers of the bushings would equal $\frac{360}{2 \times 11}$ = 16.36 degrees. The sine of 16.36 degrees is 0.2815, and 0.2815 × 8 = 2.252 inches. The arms are adjusted to locate the centers of the bushings this distance apart by placing closely fitting plugs in the bushings and measuring from one plug to another with a micrometer or vernier caliper. Of course, when taking this measurement, allowance is made for the diameter of the plugs.

After the arms are set, a hole is drilled and reamed; an accurately fitting plug is then inserted through the bushing and hole to secure the arms when drilling and reaming the adjacent hole. The radial arms are then indexed one hole so that the plug can be inserted through one of the arms and the last hole reamed. The third hole is then drilled and reamed, and this operation is repeated for all of the holes. Evidently, if the center-to-center distance between the bushings is not exactly right, the error will be indicated by the position of the arms relative to the last hole and the first one reamed; moreover, this error will be multiplied as many times as there are holes. For instance, if the arms were too far apart, the difference between the center-to-center distance of the last pair of holes and the center-to-center distance of the bushings in the arms would equal, in this particular case, eight times the error, and the arms should be re-adjusted accordingly. Larger bushings would then be inserted in the arms and the holes re-reamed, this operation being repeated until the holes were all equi-distant.

As will be seen, the value of this method lies in the fact that it shows the accumulated error. Thus, if the arms were 0.0005 inch too far apart, the difference between the first and last hole would equal 8 × 0.0005 = 0.004 inch. This same principle of dividing can be applied in various ways. For instance, the radial arms, if slightly modified, could be used for drilling equally-spaced holes in the periphery of a plate or disk, or, if a suitable marking device were attached, a device of this kind could be used for accurately dividing circular parts.

Originating a Precision Dividing Wheel. — There are various methods employed for making accurate indexing wheels for a definite number of divisions. One of these methods, suitable particularly for small numbers of divisions, employs a split wheel with a series of taper holes reamed through the two divisions. By shifting the two divisions from point to point (as explained in connection with sketch *B*, Fig. 16) and reaming and re-reaming the taper holes at each shifting, they may finally be brought very accurately into position. Another method that has been employed consists in clamping about the rim of the dividing wheel a number of precisely similar blocks, fitting close to each other and to the wheel itself. These blocks are then used for locating the wheel in each of its several positions in actual work. A third and simpler method (a modification of the one last described) consists in grinding a series of disks and clamping them around a rim of such diameter that the disks all touch each other and the rim simultaneously, as explained in connection with sketch *C*, Fig. 16. The wheel described in the following, which is illustrated in Fig. 17, was made in this way.

Disks *A* are clamped against an accurately ground face of the wheel *B* and are supposed to just touch each other all around, and to be each of them in contact with the ground cylindrical surface at *x*. They are held in proper position by bolts *C* and nuts *D*. The bolts fit loosely in the holes of the disks or bushings *A* so that the latter are free to be located as may be desired with reference to the bolts.

One or two improvements in the construction of this type of dividing wheel may be noted before proceeding to a description of the way in which it is made. For one thing, instead of having an indexing bolt enter the V-space between two adjoining disks, a smaller diameter *y* is ground on each of them, over which locking finger or pawl passes, holding the wheel firmly from movement in either direction. This construction has the advantage of a probable lessening of error by locating on each bushing instead of between two bushings; moreover, it gives a better holding surface and better holding angles than would be the case if this smaller diameter were not provided.

A second improvement lies in the method of clamping the bushings *A* in place.   Instead of providing each bolt with a separate washer, a ring *F* is used.   This ring fits closely on a seat turned

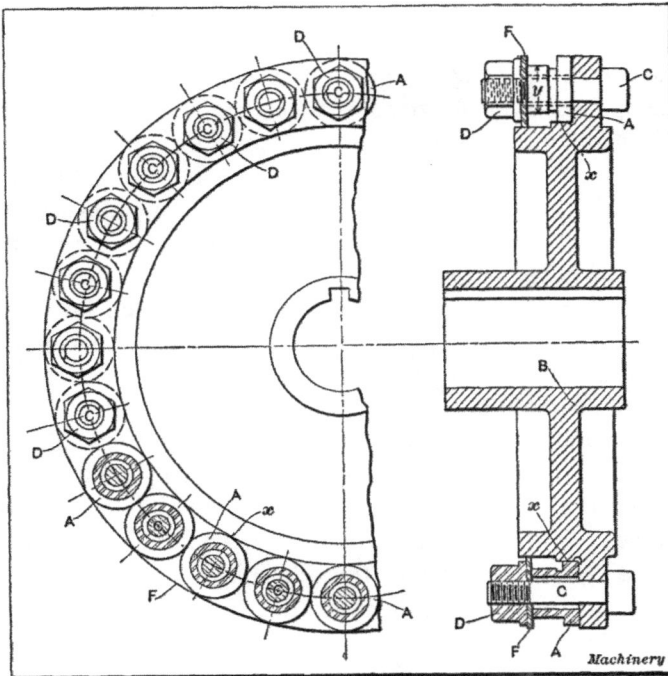

Fig. 17.   A Precision Dividing Wheel

to receive it on the dividing wheel *B*.   When one bushing *A* has been clamped in place, the disk is locked from movement so that there is no possibility, in clamping the remaining bushings, of having their location disturbed in the slightest degree by the turning of the nuts in fastening them in place.

The bushings *A*, of which there were in this case 24, were first all ground exactly to the required diameters on their locating and locking surfaces.   The important things in this operation are, first, that the large or locating diameter of the bushing should be exactly to size; and second, that this surface should be in exact alignment with the diameter in which the locking is done; and, finally, that the face of the bushing should be squared with the cylindrical surfaces. A refined exactness for the diameter of the locking surfaces is not so important, as the form of locking device provided allows slight variations at this point without impairment of accuracy. This dimension was kept within very close limits, however. The truth of the two cylindrical surfaces and the face of the bushing was assured by finishing all these surfaces in one operation on the grinding machine.

The sizing of the outer diameter of the bushing, which is 1.158 inch, had to be so accurate that it was not thought wise to trust to the ordinary micrometer caliper. An indexing device was, therefore, made having a calipering lever with a long end, in the ratio of 10 to 1, which actuated the plunger of a dial test indicator of the well-known type made by the Waltham Watch Tool Co. The thousandth graduations on the dial of this indicator would then read in ten-thousandths, permitting readings to be taken to one-half or one-quarter of this amount. The final measurements with this device were all taken after dipping the bushings in water of a certain temperature, long enough to give assurance that this temperature was universal in all the parts measured. It will be understood, of course, in this connection, that getting the diameter of these bushings absolutely to 1.158 inch was not so important as getting them all exactly alike, whether slightly over or slightly under this dimension; hence, the precaution taken in measurement.

Wheel B was next ground down nearly to size, great care being taken that it should run exactly concentric with the axis. As soon as the diameter of the surface x was brought nearly to the required dimension as obtained by calculation, the disks were tried in place. The first one was put in position with its loose hole central on the bolt and clamped in place under ring F. The next bushing was then pressed up against it and against the surface x of the wheel and clamped in place. The third one was similarly clamped in contact with its neighboring bushing and the wheel, and so on, until the whole circle was completed.   It was then found that the last disk would not fill the remaining space. This required the grinding off of some stock from surface x, and a repetition of the fitting of the bushings A until they exactly filled the space provided for them.

This operation required, of course, considerably more skill than a simple description of the job would indicate.   One of the points that had to be carefully looked out for was the cleaning of all the surfaces in contact.   A bit of dust or lint on one of the surfaces would throw the fitting entirely out.   The temperature of the parts was another important consideration.   As an evidence of the accuracy with which the work was done, it might be mentioned that it was found impossible to do this fitting on a bench on the southern or sunny side of the shop, the variations of temperature between morning and noon, and between bright sunshine and passing clouds, being such that the disks would not fit uniformly.   The variation from these minute temperature changes resulted from the different coefficients of expansion of the iron wheel and the steel bushings.   The obvious thing to do would be to build a room for this work kept at a constant temperature and preferably that of the body, so that the heat of the body would make no difference in the results.   It was found sufficient in this case, however, to do the work on the northern side of the shop where the temperature was more nearly constant, not being affected by variations in sunshine.

Generating a Large Index Plate. — An index plate was required for cutting some large gear rings, 10, 12 and 14 feet in diameter, respectively.   These gears, which were for driving vertical boring mills, had to be cut very accurately as they were to be driven by two pinions located diametrically opposite each other.   Such a drive requires much greater accuracy in the gear than when there is only one driving pinion.   Suppose a gear has 176 teeth and that every pitch is 0.001 inch too large over the first half of the gear, and 0.001 inch too small over the last half; this error would be so small that there would be no appreciable defect in the drive, so long as there is only one pinion; but space No. 89 would be 0.088 inch out of place, and, when there are two pinions, it would be this space which would be acted upon by the second pinion, while the first pinion engages space No. 1. Now, 0.088 inch is entirely too much of an error to be neglected; in fact, such an error would condemn the whole drive, and yet, the individual error in the spaces would be so small as to make it almost impossible to avoid them. It is this possibility of errors accumulating which makes it so difficult to originate a correct dividing plate. The following method of generating a large index plate was described by A. L. DeLeeuw in the August, 1905, number of MACHINERY.

The gears to be cut were to have 150, 176 and 210 teeth, respectively, of $1\frac{1}{4}$ diametral pitch. They were to be cut on the slotter, as there was no gear cutter large enough to handle such work. The section of the gear was like that shown at A, Fig. 18, the fit being at a. A number of bolts were to secure the gear to the table of the boring mill which it was to drive. The ordinary way to cut a gear of this type was to turn it up and fit it to a spider or index plate. The bolt holes in this spider were used as jig holes to drill the holes in the rim of the gear, thus making all gears interchangeable. The spider had notches milled in its outer circumference. This was done on the gear cutter, with the result that if the gear cutter had any errors, those errors would be reproduced in the spider. As a rule, the diameter of the spider was very much larger than the diameter of the dividing worm-wheel on the gear cutter, so that the errors of the worm-wheel were not only copied, but even enlarged. All this was absolutely inadmissible on this job, so new spiders were made and provided with as many holes as the gear was to have teeth. These holes were all drilled at equal distances from the center, and very accurately spaced; and the spacing and drilling were done in such a manner that the errors could not accumulate.

Dividing Mechanism for Generating Index Plate. — The plate or spider to be divided was first turned true and with a smooth surface J, Fig. 18, on the side to be used for dividing. The gear is shown in place by the dotted lines although it was not attached to the spider while the spacing was being done. The spider was pivoted on a stud B held in a stand. The outer rim of the spider

was supported by numerous jack-screws, with brass caps, allowing the spider to slide over the jacks, but preventing it from sag-
ging. The jack-screws were locked in position, as shown. On
stud *B* two arms *C* and *D* were mounted as shown by the plan
view. These arms are about 7 feet 8 inches long from the center
of the stud to the outer end. They are cast iron, and were made
as light as was consistent with the required stiffness. Both arms
were free to turn around the stud *B*. The bearings on the stud

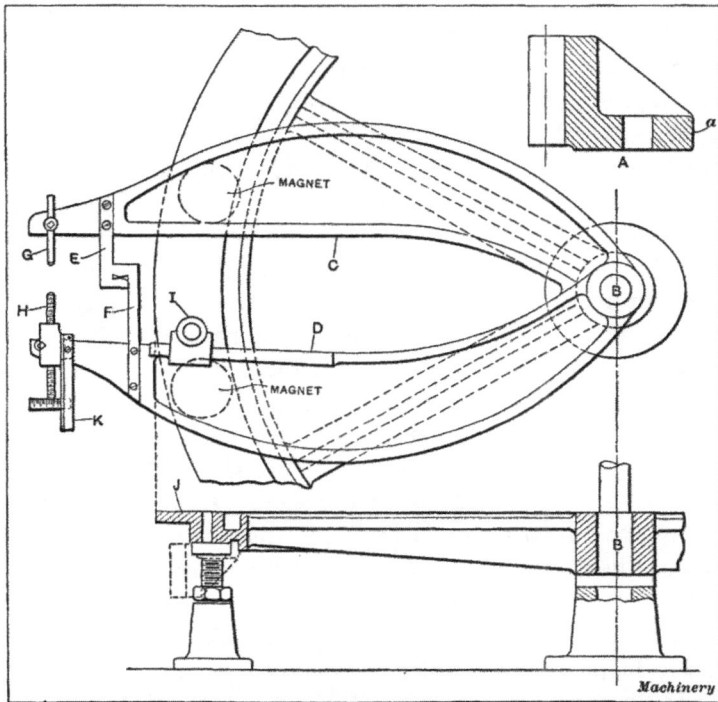

Fig. 18. Dividing Mechanism for Generating a Large Index Plate

were, of course, one above the other, but the main part of the
arms were in the same plane. When in position they formed a
pair of large jaws. A hardened steel hook *E* and another hook *F*
were fitted to the end of each arm, and these hooks coming in
contact when the arms were a certain distance apart made this
distance the maximum space between them.

The arm C is provided at its extreme end with a piece holding a hardened steel pin G rounded at its end. This piece could be swiveled so as to allow the pin to point in various directions. When once adjusted, the swiveling piece was clamped in position. The other arm D had a similar swiveling piece which was split and provided with a clamping screw. It was tapped out and held a micrometer screw H, which could be clamped by means of the split bearing and clamping screw. The end of the screw H was rounded over, and the head was made large and divided so that one division corresponded to 0.0005 inch. No attempt was made to have this screw or the divisions on the head very accurate, for, as will be seen later, great accuracy would serve no purpose. A pointer K was fastened to the swiveling piece. This pointer was simply a piece of sheet steel with beveled edge, and long enough to always be over the screw head. The piece G and the screw H were adjusted so as to be in line when against each other. In this manner, the movement of the arms relative to each other was limited in both directions; one way by the hooks E and F and the other way by the parts G and H.

Operation of Precision Dividing Mechanism. — The method of operating this spacing or dividing mechanism shown in Fig. 18 was as follows: One arm was pulled back until the hooks came in contact, when it was clamped to the spider. Then the other arm was moved up until the screw H struck the pin G; this arm was then clamped and the other released. The first arm was then pulled back again, and these operations were repeated, thus moving both arms all around the spider. The arm D was provided with a planed edge, to which a piece I was securely fastened. This piece is shown in detail at A, in Fig. 19. It consisted of a slide or jig having a bushed hole to be used later for drilling the holes. It also carried a brass block l which was carefully scraped to a bearing on the spider. To the block was screwed a glass-hard steel piece m having a beveled edge, and located about 0.001 inch above the spider. This steel piece formed a scribing edge for marking lines on the spider. A line was marked on the spider and the screw set so as to get approximately the correct division. The arms were then brought around as described, and, after the required number of spaces were made, the distance from the scratch edge to the starting line was noted. This would give some idea as to how much correction was to be made, and the correction was made by means of the micrometer screw. The whole method was one of trial, and practically the same as spacing with a pair of dividers; except that where dividers fail, this method succeeds. Any one who ever tried to space with dividers knows how impossible it is to get the same result twice. The point slips off in some minute hole of the iron, or the angle at which the dividers are held varies, or some other little thing happens, which makes the

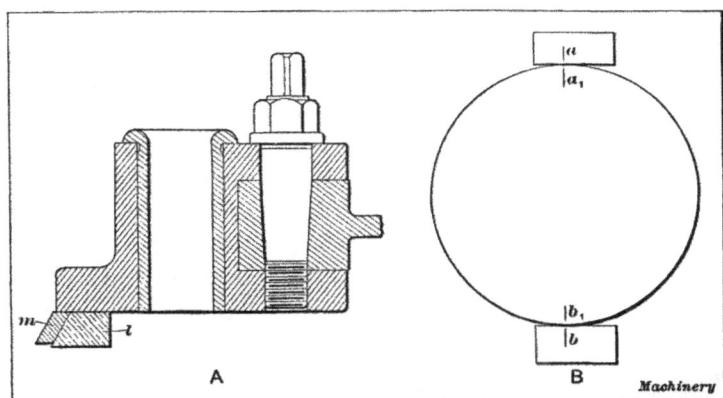

Fig. 19. (A) Jig and Scribing Block shown at I, Fig. 18. (B) Diagram Illustrating Preliminary Method of Dividing Index Plate

result unreliable. If one succeeds in setting the dividers so as to come back to the starting point, say, after 176 steps or divisions, this same setting will give entirely different results if the same man goes around once more.

The clamping of the arms C and D, Fig. 18, is accomplished by means of electro-magnets fastened to the under side of the arms. The magnets were wound so that they would stay cool on a 220-volt circuit, when in series with a 110-volt 16-candle-power lamp. This lamp acted as pilot lamp, indicating when the current was turned on, and also indicating when something was wrong with the winding. It is obvious that holding the arms in position by means of magnets is far superior to clamping, as the latter method would easily throw them out of square. Even the magnets were liable to cause trouble. If the surface of the plate is not perfectly level, smooth, true, well supported at frequent intervals, and on a rigid foundation where it is not disturbed by vibrations of nearby machines, and if the magnets are not scraped nicely to a bearing with the plates and also to the arms, and if they are not protected so that no dirt can creep between them and the plates, and if one of several other things is not provided for, there is trouble; and the difficulty is that this trouble does not become evident until one has gone clear around the circle, and even then it may not be discovered immediately. One may find an accumulated error of $\frac{1}{16}$ inch after going once around; after correcting for this amount, and going around once more, one may find an error in the other direction. Trial after trial may not produce any better result, and one may think that it is simply bad luck in estimating the correction to be made; whereas, in reality, it is simply the result of poor working conditions. Great care and considerable patience are necessary, but with those elements at hand, the method gives fine results.

To start the spacing, two lines diametrically opposite each other were marked on the plate. To get these lines correct, the following method was employed: Two stations were erected approximately at opposite ends of a diameter of the plate. These stations were merely castings blocked up to the proper height, so as to bring their top surfaces level with the top of the plate. A piece of sheet brass was fastened to one of the stations, whereas the other station received an adjustable piece of sheet brass. Lines were drawn on these pieces of brass so as to be nearly in line with each other and with the center of the spider. This, of course, was merely guesswork. These lines were extended onto the plate, so that the lines appeared as shown by the diagram B, Fig. 19. The two lines on the brass are marked a and b; and the two extensions of these lines on the plate a_1 and b_1. The plate was now turned until b_1 came exactly opposite a. Now, if the lines a_1 and b_1 are exactly diametrically opposite each other, then the line a_1 is bound to come exactly opposite the line b. This, of course, was found not to be the case. The distance between a_1 and b was carefully divided into two equal parts, and a new line a_1 and b (one an extension of the other) was drawn on the brass and on the plate. Then another half turn was given to the plate, and the line a_1 was brought opposite the line a, and the error (if any) noticed. This was repeated until no further error was discovered. The object of dividing the plate into two equal parts was simply to facilitate the first attempts at spacing, as an appreciable error could be discovered by going only halfway around. The lines a_1 and b_1 were not used, however, for the final spacing.

The approximate distance was measured from the center of the stud B, Fig. 18, to the center of the micrometer screw H. From this measurement, the circumference of the circle, described by the screw point, was computed. This was divided by, say, 176, and the quotient was the distance from the point of the hardened plug to the point of the micrometer screw. A pair of inside calipers was then used to set the screw to this distance. Then, 88 such distances were spaced, and the result noted. Suppose the scratch block was $\frac{3}{8}$ inch from the line b_1, Fig. 19; then an adjustment was made of $\frac{1}{88}$ part of $\frac{3}{8}$ inch, or about $4\frac{1}{4}$ thousandths of an inch, which corresponds to about $8\frac{1}{2}$ divisions of the micrometer screw head. It would have been a difficult matter to estimate the amount of correction to be made, if not assisted by the micrometer screw. One must not expect to secure the correct adjustment the first time, but considering the difficulty of the job it is surprising how few trials are necessary to get perfect spacing. It never took more than one day (of ten hours) to get a perfect spacing, after everything was arranged. When the scratch block comes to the edge of the first line, after the required number of steps, the spacing is considered finished. It remains, however, to prove that this result was not obtained by some hidden defect; therefore, the spacing was checked by a second trial. If the scratch block came against the outside edge of the first line again, it was considered safe to proceed with the next operation of drilling the holes.

Method of Drilling Equally-spaced Holes in Index Plate. — Before drilling the holes, the spacing arms C and D, Fig. 18, were indexed around again, and a line drawn along the scratch block (attached to jig I) at every step or division, thus marking a line for every hole to be drilled. This operation was an additional check on the accuracy of the spacing. As there were to be 210 holes in the largest spider, stepping around three times (once for the last trial, once for checking and once for marking the lines) required 630 spaces. The lines drawn were, at the most, 0.002 inch wide. A magnifying glass was used, to see if the scratch block came up against the line; so that it is safe to say the total error was less than 0.002 inch; hence, the error in the individual spaces was an exceedingly small amount.

One of the two arms was now removed from the spider, the arm with the drill jig remaining. A radial drill had previously been put in place, and the dividing arm was secured against the radial drill column. The spider was rotated under the arm by means of block and tackle, so that the consecutive lines coincided with the scratch block. A magnifying glass was used to locate the spider accurately. The hole was then drilled $\frac{1}{32}$ inch small. The holes were to be 1-inch ultimately, so that a $\frac{31}{32}$ inch drill was used. The drill was then replaced by a reamer 0.002 inch below size, and the reaming was done by power, the drill jig bushing being replaced by a new one corresponding in size with the reamer. After this operation was completed, the jig bushing was removed and a new one inserted, having a bore exactly 1 inch in diameter. This jig was used for reaming by hand.

One might naturally wonder why the second arm was removed and a different method used for drilling than for spacing. There were two reasons why this was done. The first one is that, when drilling by the spacing method, the spindle of the drill must follow the arms which would occupy different positions on the spider; hence, the arms and spider would have to be shifted under the drill spindle for each hole, or, in the case of a radial drill being used, the drill spindle might be aligned with the jig. This would mean very careful adjustment repeated many times. By leaving the drill spindle stationary, and one arm blocked against the column, the relative adjustment of drill spindle and the jig is preserved. The second reason is of an entirely different nature. Electro-magnets were used for the purpose of holding the arms in place, as previously mentioned. This proved to be an excellent method; in fact, it would have been absolutely impossible to hold the arms without springing, and with uniform pressure by other means; but, when drilling, chips would crowd in between the magnet and the spider, in spite of all precautions, even though the plate was cleaned after drilling a hole. These little chips would tilt the arm forward and crowd the drill. For these reasons, the magnets were not used while drilling, the clamping being done by an ordinary C-clamp. It is true that the setting of the arm was not so perfect as when done by means of the spacing method, but it was good enough for the purpose, for it must not be forgotten that it was not the object to get a gear with absolutely even pitches, but one with reasonably correct pitches, and without an accumulation of errors. This latter requirement was taken care of by the method of spacing; a good mechanic can set an edge so close to a fine line that the error is not appreciable in a gear of $1\frac{1}{4}$ diametral pitch.

Application of the Large Index Plate. — It may be of interest to know something about how the large index plate, generated as described in the foregoing, was used when cutting large gears. The gear cutting was done on a slotter, specially designed and used for cutting internal and external gears which are too large in diameter or of too great a pitch to be cut on an ordinary gear cutter. The table is very long and in the shape of a cross, thus giving a large surface for the support of the spiders, without making the table too heavy. For unusually large gears additional supports are provided, in the nature of stands placed around the machine, each on its own foundation. A T-slot runs the long way of the table and exactly in line with the center-line of the ram. A stud and bushing held in this T-slot, serve as pivot for the spiders. The lower end of the ram has a square hole to take the tools which are placed in a horizontal position. The shank of the tool is planed to a gage so that the center-line of the tool has always the same distance from the planed side of the shank. Inserting the tool in the square hole, and forcing it up by means of set-screws against one side of this hole, brings the center-line of the tool in line with the center-line of the machine, and thus with the center-line of the gear to be cut. The gear blank is fed toward the tool until within one or two hundredths of an inch of the required depth, when the power feed is thrown out, and the blank is further fed by hand. A stop prevents the operator from making a mistake as to depth. In addition, several other precautions are taken. The pitch line and the bottom line of the teeth are marked on the blank by means of a scratch gage. In case the teeth are blocked out in the casting (for the purpose of securing a better casting, at least around the teeth) the operator first goes once around, outlining the tooth shape on the blank, to make sure that there is sufficient stock for every tooth. In order to do this outlining he brings the tool down to within $\frac{1}{32}$ inch from the blank, and marks around it with a scratch awl. A screw is used to move the blank when it is unusually heavy, otherwise the spacing is done by hand. For indexing the gear the holes in the index plate were engaged by a pin located under the spider and held by a suitable bracket. The pin was withdrawn by means of a lever.

The gears proved to be accurately spaced, and the two driving pinions showed a perfect bearing on the gear teeth. Though the side clearance in the teeth was only 0.01 inch, such a thing as a bearing on the back of a tooth never occurred.

Locating Work for Boring on Milling Machine. — It is often desirable to perform boring operations on the milling machine, particularly in connection with jig work. Large jigs, which because of their size or shape could not be conveniently handled in the lathe, and also a variety of smaller work, can often be bored to advantage on the milling machine. When such a machine is in good condition, the necessary adjustments of the work in both vertical and horizontal planes can be made with considerable accuracy by the direct use of the graduated feed-screw dials. It is good practice, however, when making adjustments in this way, to check the accuracy of the setting by measuring the center distances between the holes directly. For the purpose of obtaining fine adjustments when boring on the milling machine, the Brown & Sharpe Mfg. Co. makes special scales and verniers that are attached to milling machines, so that the table may be set by direct measurement.   By attaching a scale and vernier to the table and saddle, respectively, and a second scale to the column with a vernier on the knee, both longitudinal and vertical measurements can be made quickly and accurately, and the chance of error resulting from inaccuracy of the screw, or from lost motion between the screw and nut, is eliminated.

Checking Location of Holes by Micrometer-and-plug Method. — One method of checking the accuracy of the location of holes bored in the milling machine, is to insert closely fitting ground plugs into the bored holes and then determine the center-to-center distance by taking a direct measurement across the plugs with a micrometer or vernier caliper.   For example, if holes were to be bored in a jig-plate, as shown in Fig. 1, assuming that hole A were finished first, the platen would then be moved two inches, as shown by the feed dial; hole B would then be bored slightly under size.   Plugs should then be accurately fitted to these holes, having projecting ends, preferably of the same size. By measuring from one of these plugs to the other with a vernier or micrometer caliper, the center distance between them can be accurately determined, allowance being made, of course, for the radii of each plug.   If this distance is incorrect, the work can be adjusted before finishing B to size, by using the feed-screw dial.   After hole B is finished, the knee could be dropped 1.5 inch, as shown by the vertical feed dial, and hole C bored slightly under size; then by the use of plugs, as before, the location of this hole could be tested by measuring center distances between C—B and C—A.

An example of work requiring the micrometer-and-plug test is shown set up in the milling machine in Fig. 20.   The large circular plate shown has a central hole and it was necessary to bore the outer holes in correct relation with the center hole within a limit of 0.0005 inch.   The center hole was first bored and reamed to size; then an accurately fitting plug was inserted and the distances to all the other holes were checked by measur-

ing from this plug. This method of testing with the plugs is intended to prevent errors which might occur because of wear in
the feed-screws or nuts, that would cause the graduated dials to
give an incorrect reading. On some jig work, sufficient accuracy
could be obtained by using the feed-screw dials alone, that is,
without testing with the plugs, in which case the accuracy would
naturally depend largely on the condition of the machine.

Combination Boring Tool and Test Plug. — A method that is a
modification of the one in which plugs are used to test the center

Fig. 20. Example of Precision Boring on Milling Machine

distance is as follows: All the holes are first drilled with suitable allowance for boring, the location being obtained directly
by the feed-screw dials. A special boring-tool, the end of which
is ground true with the shank, is then inserted in the spindle and
the first hole, as at *A* in Fig. 1, is finished, after which the platen
is adjusted for hole *B* by using the dial as before. A close-
fitting plug is then inserted in hole *A* and the accuracy of the
setting is obtained by measuring the distance between this plug
and the end of the boring-tool, which is a combination tool and

test plug. In a similar manner, the tool is moved from one position to another, and, as all the holes have been previously drilled, all are bored without removing the tool from the spindle.

Locating Work from Turned Plug in Spindle Chuck. — Another modification of the micrometer-and-plug method is illustrated in Fig. 21. It is assumed that the plate to be bored is finished on the edges, and that it is fastened to an angle-plate, which is secured to the table of the milling machine and set square with

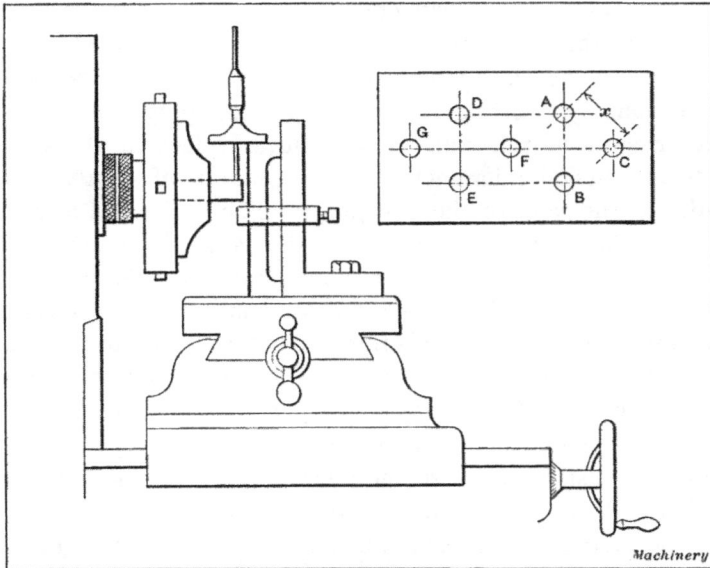

Fig. 21. Obtaining Vertical Adjustment by Means of Depth Gage and Turned Plug in Chuck

the spindle. A piece of cold-rolled steel or brass is first fastened in the chuck (which is mounted on the spindle) and turned off to any diameter. This diameter should preferably be an even number of thousandths, to make the calculations which are to follow easier. The turning can be done either by holding the tool in the milling machine vise, or by securing it to the table with clamps. In either case, the tool should be located near the end of the table, so as to be out of the way when not in use.

After the piece in the chuck is trued, the table and knee are adjusted until the center of the spindle is in alignment with the

center of the first hole to be machined. This setting of the jig-
plate is effected by measuring with a micrometer depth gage
from the top and sides of the work, to the turned plug, as shown
in the illustration. When taking these measurements, the radius
of the plug in the chuck is, of course, deducted. When the plate
is set the plug is removed from the chuck and the first hole
drilled and bored or reamed to its proper size. We shall assume
that the holes are to be located as shown by the detail view,
and that hole A is the first one bored. The plug is then again
inserted in the chuck and trued with the tool, after which it is
set opposite the place where the second hole B is to be bored;
this is done by inserting an accurately fitting plug in hole A and
measuring from this plug to the turned piece in the chuck,
with an outside micrometer. Allowance is, of course, again
made for the radii of the two plugs. The horizontal measurement can be taken from the side of the work with a depth gage
as before. The plug is then removed and the hole drilled and
bored to the proper size. The plug is again inserted in the chuck
and turned true; the table is then moved vertically to a position
midway between A and B, and then horizontally to the proper
position for hole C, as indicated by the depth gage from the
side of the work. The location can be verified by measuring
the center distances x with the micrometer. In a similar manner holes D, E, F and G are accurately located.

If the proper allowances are made for the variation in the size
of the plug, which, of course, is made smaller each time it is
trued, and if no mistakes are made in the calculations, this method
is very accurate. Care should be taken to have the gibs on all
slides fairly tight at the beginning, and these should not be tightened after each consecutive alignment, as this generally throws
the work out a few thousandths. If the reductions in the size of
the plug, each time it is turned, are confusing, new plugs can be
used each time a test is made, or the end of the original plug can
be cut off so that it can be turned to the same diameter for every
test. If the center distances x are not given, it is, of course, far
more convenient to make all the geometric calculations before
starting to work.

The Button-and-plug Method. — The use of the button method as applied to the milling machine, is illustrated in Fig. 22, where a plain jig-plate is shown set up for boring. The jig, with buttons B accurately located in positions corresponding to the holes to be bored, is clamped to the angle-plate A that is set at right angles to the spindle. Inserted in the spindle there is a plug P, the end of which is ground to the exact size of the indicating buttons. A sliding sleeve S is accurately fitted to this plug and when the work is to be set for boring a hole, the table and knee of the

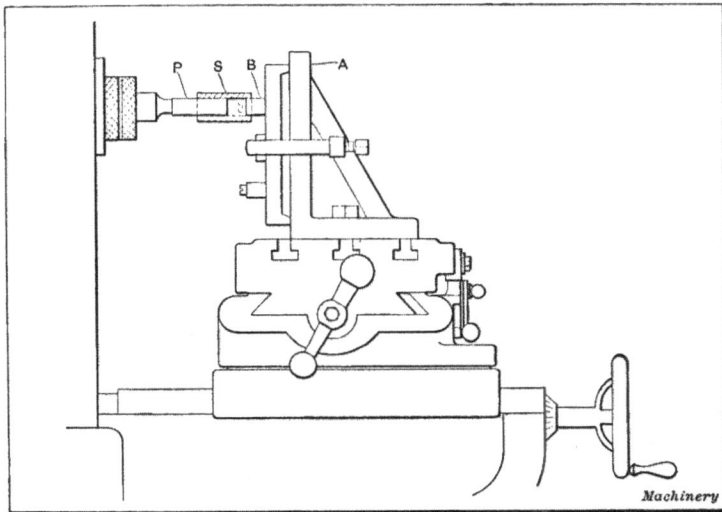

Fig. 22. Accurate Method of Aligning Spindle with Button on Jig Plate

machine are adjusted until the sleeve S will pass over the button representing the location of the hole, which brings the button and spindle into alignment. When setting the button in alignment, all lost motion or backlash should be taken up in the feed-screws. For instance, if the button on the jig should be a little higher than the plug in the spindle, do not lower the knee until the bushing slips over the button, but lower the knee more than is required and then raise it until the bushing will pass over the button. This same rule should be followed for longitudinal adjustments.

After the button is set by this method, it is removed and the plug in the spindle is replaced by a drill and then by a boring tool or reamer for finishing the hole to size. In a similar manner the work is set for the remaining holes. The plug P for the spindle must be accurately made so that the outer end is concentric with the shank, and the latter should always be inserted in the spindle in the same relative position. With a reasonable degree of care, work can be set with considerable precision by this method, providing, of course, the buttons are properly set.

Some toolmakers use, instead of the plug and sleeve referred to, a test indicator for setting the buttons concentric with the ma-

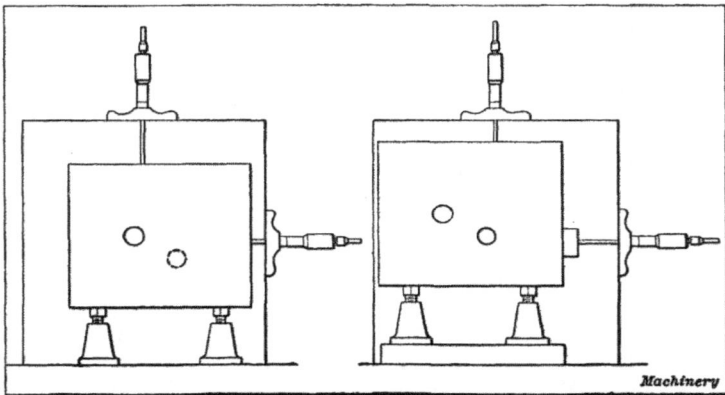

Fig. 23.   Locating Work from Edges of Angle Plate by Means of Depth Gages and Size Blocks

chine spindle. This indicator is attached to and revolves with the spindle, while the point is brought into contact with the button to be set. The difficulty of seeing the pointer as it turns is a disadvantage, but with care accurate results can be obtained.

Size Block and Gage Method. — Another method which can at times be employed for accurately locating a jig-plate in different positions on an angle-plate, is shown in Fig. 23. The angle-plate is, of course, set at right angles to the spindle and depth gages and size blocks are used for measuring directly the amount of adjustment. Both the angle-plate and work should have finished surfaces on two sides at right angles to each other, from which measurements can be taken. After the first hole has been bored, the plate is adjusted the required distance both horizontally and vertically, by using micrometer depth gages, which should preferably be clamped to the angle-plate. If the capacity of the gages is exceeded, measurements may be taken by using standard size blocks in conjunction with the depth gages.

It is frequently necessary to bore holes in cast jig-plates or machine parts, which either have irregularly shaped or unfinished edges. A good method of locating such work is illustrated in Fig. 24. The part to be bored is attached to an auxiliary plate *A* which should have parallel sides and at least two edges which

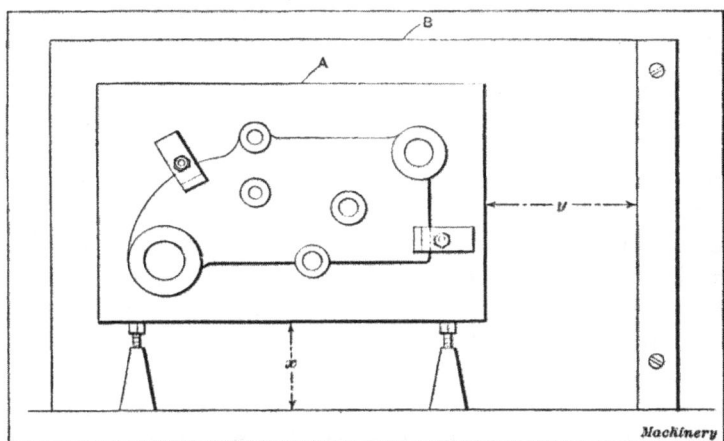

Fig. 24. Method of Holding and Locating Casting of Irregular Shape,
for Boring Holes

are straight and at right angles to each other. This auxiliary plate with the work is clamped against an accurate angle-plate *B*, which should be set square with the axis of the machine spindle. A parallel strip is bolted to the angle-plate and the inner edge is set square with the machine table. After the first hole is bored, the work is located for boring the other holes, by taking vertical measurements x from the table to the edge of the auxiliary plate, and horizontal measurements y between the parallel and the plate. These measurements, if quite large, might be taken with micrometer gages, whereas, for comparatively small adjustments, size blocks might be more convenient.

Vernier Height Gage and Plug Method. — When a vernier height gage is available, it can often be used to advantage for setting work preparatory to boring in a milling machine. One advantage of this method is that it requires little in the way of special equipment. The work is mounted on an angle-plate or directly on the platen, depending on its form, and at one end an angle-plate is set up with its face parallel to the spindle. An accurately finished plug is inserted in the spindle and this plug is set vertically from the platen and horizontally from the end angle-plate, by measuring with the vernier height gage. After the plug is set for each hole, it is, of course, removed and the hole drilled and bored or reamed.

The way the plug and height gage is used is clearly illustrated in Figs. 25 and 26. The work, in this particular case, is a small jig. This is clamped directly to the machine table and at one end an angle-plate is also bolted to the table. This angle-plate is first set parallel with the traverse of the saddle or in line with the machine spindle. To secure this alignment, an arbor is inserted in the spindle and a test indicator is clamped to it by gripping the indicator between bushings placed on the arbor. The table is then moved longitudinally until the contact point of the indicator is against the surface plate; then by traversing the saddle crosswise, any lack of parallelism between the surface of the angle-plate and the line of saddle traverse will be shown by the indicator.

When the work is to be adjusted horizontally, the vernier height gage is used as shown in Fig. 26, the base of the gage resting on the angle-plate and the measurement being taken to an accurately ground and lapped plug in the spindle. For vertical adjustments, the measurements are taken between this ground plug and the machine platen as in Fig. 25.

Locating Holes to be Bored from Center-punch Marks. — The problem of accurately locating holes to be bored on the milling machine has received much attention, and the method generally used when accuracy has been required is the button method, which was previously described. So much time is required for doing the work by this method, however, that numerous efforts have been made to obtain equally good results in other ways.

Fig. 25. Making a Vertical Adjustment by Measuring to Ground
Plug in Spindle

Fig. 26. Making a Horizontal Adjustment by Measuring from Angle
Plate to Ground Plug

The increasing demand for rapidity combined with accuracy and a minimum liability of error, led to the development of the system described in the following: A center-punch mark takes the place of the button, by which to indicate the proper position of the work for boring. The fundamental principle involved is to lay out, accurately, two lines at right angles to each other, and correctly center-punch the point where they intersect. With proper care, lines may be drawn with a vernier height gage at right angles, with extreme accuracy, the chief difficulty being to

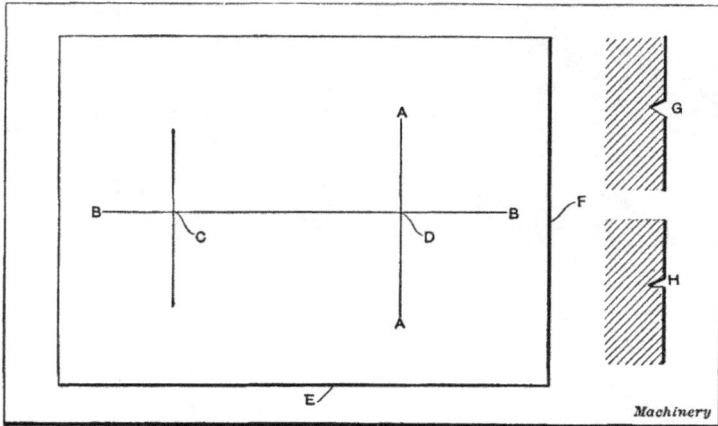

Fig. 27.   Diagram Illustrating Rapid but Accurate Method of Locating Holes to be bored on Milling Machine

accurately center the lines where they cross. For semi-accurate work this may be done with a common center-punch, but where extreme accuracy is required this method is not applicable, as the average man is incapable of marking the point of intersection accurately.

The diagram, Fig. 27, illustrates, in a simple way, the procedure adopted in laying out work by this system. The base E is in contact with a surface plate while the line BB is drawn with a height gage; then with side F on the plate the line AA is drawn. It will be seen that these lines will be at right angles to each other, if the bases E and F are square. Work done by this method must have two working surfaces or base lines, and these must be at right angles to each other. There is no difficulty in drawing the locating lines *AA* and *BB* correctly, either with a vernier height gage or with a special micrometer gage reading to 0.0001 inch, the only difficult element being to accurately center-punch the lines where they intersect as at *D*.   It is assumed that two holes are to be bored, so that the intersection at *C* would also be center-punched.

The scriber point of the height gage should be ground so that it will make a V-shaped line, as shown by the enlarged sketch *G*, rather than one which would resemble a saw-tooth, as at *H*, if a

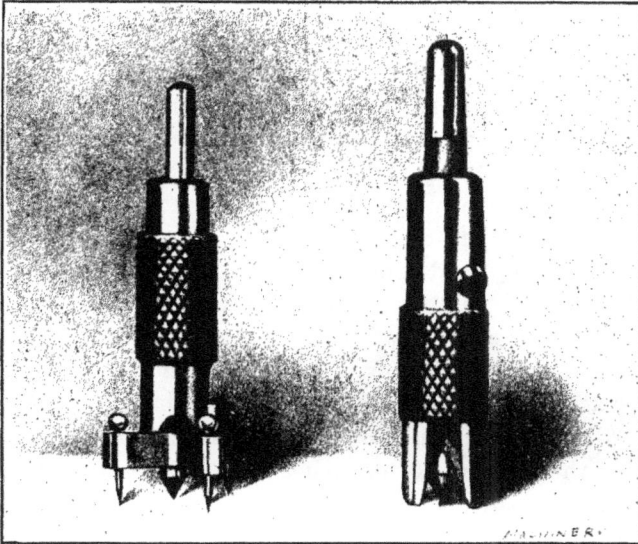

Fig. 28.   Center Locating Punch   Fig. 29.   Center Enlarging Punch

cross-section of it were examined with a microscope.   This is important because when the lines are V-shaped, an accurate point of intersection is obtained.

As it is quite or almost impossible to accurately center-punch the intersection of even two correctly drawn lines, by ordinary means, the punch shown in Figs. 28 and 30 was designed and an extended experience with it on a very high grade of work has demonstrated its value for the purpose.   It consists essentially of a small center-punch *O* (Fig. 30) held in vertical position by a holder *P* which is knurled to facilitate handling.   Great care

should be exercised in making this tool to have the body of the
punch straight, and to have it stand at right angles to the surface
to be operated upon, for the slightest inclination will cause the
finished hole to be incorrect, no matter how carefully the lines
are drawn.   The 60-degree point must be ground true with the
axis.   The holder for the punch stands on three legs, located as
indicated, and ground accurately to a taper fit in the holder,
where they are secured by watch screws bearing on their tops.
The lower ends are hardened and terminate in an angular point
of 55 degrees (the point of the vernier scriber being 60 degrees).
The edges are sharp, and slightly rounded at the ends, so that the
legs will slide along a line smoothly.   The points V and U (Fig.
30) have edges that are in line with each other, while the point T
has an edge at right angles to the other two.   The center of the
punch is located at equal distances from all the legs, and is held
off the work normally by a leather friction acted upon by a set-
screw in the side of holder P.

If this tool is placed upon lines of the form shown at G, Fig. 27,
the legs V and U may be slid along horizontal line $B-B$, until the
sharp edge of leg T drops into line $A-A$.   When this occurs the
punch O is lightly tapped with a hammer, and the resulting mark
will be accurately located in the center of the intersection of the
lines.   It is good practice to make the work very smooth before
drawing the lines, and after laying them out, to stone them so as
to remove the slight burr raised in drawing them.   A drop of oil
is then rubbed into the lines, and the surplus wiped off.   This
procedure permits points V and U to run very smoothly along the
line, and the burr having been removed, the edge of leg T drops
into the line very readily with a slight click.   As it is not advisable to strike punch O more than a very light blow, it marks the
work but slightly, and a more distinct indentation is made with
the follower punch shown in Figs. 29 and 31.   This punch is
made like the previous one, so that it will stand at right angles to
the work.   The sectional view (Fig. 31) shows the punch A supported by the holder E which has four legs cut away on the sides
so that the point of the punch may be seen.   When this punch is
in position, it is struck a sufficiently heavy blow to make a dis-

tinctly visible mark. The work is now ready to be placed upon the work table of the milling machine, and indicated for boring the holes, an indicator being used in the milling machine spindle.

An indicator which has been found especially valuable for this purpose is shown in Fig. 32. It is of the concentric centering type, and with it the work is brought concentric with the axis of the spindle. The arbor *I* is provided with a threaded nose on

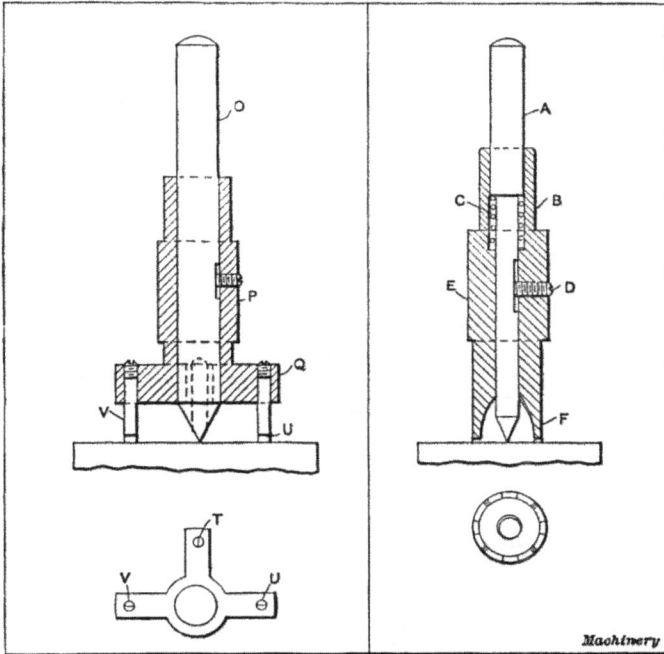

Fig. 30. Section of Center Locating Punch

Fig. 31. Section of Center Enlarging Punch

which disk *D* is screwed. This disk has four holes in its rim, equally spaced from each other. Hardened, ground, and lapped bushings *b* are put into these holes to receive plug *A* which is made a gage-fit both in these holes and in hole *B* in the outer end of sector *C*. This sector is held by a split sleeve to the barrel *L* which carries the 60-degree centering-rod *K* that comes into contact with the work to be bored. The spherical base of barrel *L* fits into a corresponding concave seat in the nose of the arbor at

H, and is held in place by a spring *E* which connects at one end
to the cylindrical stud in the base of the barrel, and at the other
to the axial rod *M* by which it and the other connecting parts may
be drawn into place, and held by the headless set-screw *J*, bearing
on a flat spot on the tang end of the rod.

Now, if plug *A* is removed from bushing *b* the point of the centering-rod *K* may be made to describe a circle. At some point
within this circle is located the center-punch mark on the work
to be bored. The holes in the rim of the faceplate all being
exactly the same distance from an axial line through both the

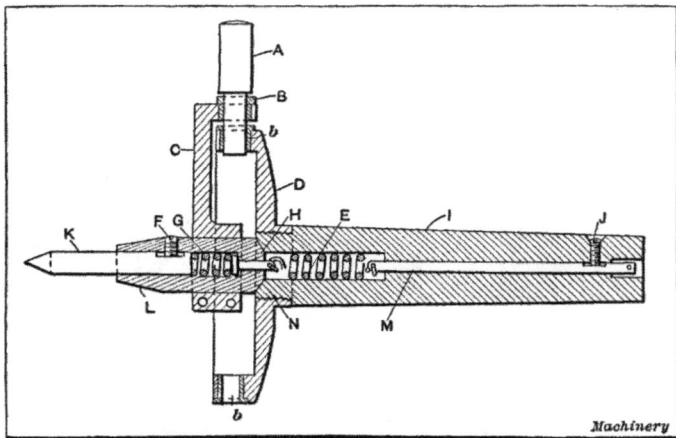

Fig. 32. Sectional View of Indicator for Aligning Center-punch Marks
Preparatory to Boring

arbor *I* and centering-rod *K*, it follows that the center mark on
the work must be so located by horizontal and vertical movements of the work table that pin *A* may be freely entered in all
the four holes in the rim of disk *D*. When that occurs, the center coincides with the axis of the spindle.

The point of the center-punch *A* (Fig. 31) should have an angle
slightly greater than the angle on the centering-rod *K*, as it is impossible to locate the work in the preliminary trials so that the
center of the work will be coincident with the axis of the spindle,
and unless the precaution mentioned is taken, the true center on
the work is liable to be drawn from its proper location when try-

ing to bring the work into such a position that the plug will enter all the holes in the disk. As the work being operated on is brought nearer to the proper location by the movements of the milling machine table, spring G will be compressed, the center rod sliding back into barrel L. This spring is made so that it will hold the center against the work firmly, but without interfering with the free rotation of the sector C around disk D. When the work is located so that the plug enters the holes, the gibs of the machine should be tightened up and the plug tried once more, to make sure that the knee of the machine has not moved sufficiently to cause the work on the table to be out of line. The work table is now clamped to prevent accidental horizontal shifting and the work is drilled and bored.

In using this indicator the milling machine spindle is not rotated together with arbor I, only the sector being turned around the disk. The tool is set, however, in the beginning, so that the axes of two of the bushings b are at right angles to the horizontal plane of the machine table, while the axes of the other holes in the disk are parallel with the top of the work table. The centering-rods are made interchangeable and of various lengths, to reach more or less accessible centers.

The principal part of the milling machine on which dependence must be placed for accuracy, when employing the method described in the foregoing, is the hole in the spindle, and this is less liable to get out of truth, from wear such as would impair the accuracy, than are the knee, table, or micrometer screws. The only moving part is the sector, and this, being light, is very sensitive. A series of 24 holes was laid out and bored in one and one-half day by the method described in the foregoing. Measurements across accurately lapped plugs in the holes showed the greatest deviation from truth to be 0.0002 inch, and running from that to accuracy so great that no error was measurable. This same work with buttons would have required considerably more time.

CHAPTER II

LAPS AND LAPPING

Lapping is a refined abrading process generally employed for correcting errors in hardened steel parts and securing a smooth surface, or for reducing the size a very small amount. Lapping bears about the same relation to the finishing of ground parts that scraping does to the finishing of planed surfaces. The lap is made of some soft metal such as cast iron or brass and it is "charged" with an abrasive which is imbedded into its surface. The grade or coarseness of the abrasive depends upon the finish required and the amount that must be removed by lapping. The form of the lap naturally depends upon the shape, size, and location of the surfaces upon which it is used.

While the main essential points of the art of lapping can be described in a book, it is necessary that the workman shall do considerable lapping before he can become proficient. There are certain motions, touches, sounds, refinements, etc., which the skilled workman acquires by practice, that are impossible to enumerate and describe in a way that would be intelligible to an inexperienced man. For instance, ask a carpenter how he knows that he is sawing a board straight, and he will be unable to tell you. Nevertheless, he has acquired a peculiar sense of touch, or such general acuteness of the senses, that he knows when the saw starts to "run out." His mind and arm automatically return the saw to a straight line without missing a stroke. It is the same way with a diemaker. He can file a die, looking only at the surface line, and can detect the instant when his file "rocks" from a straight line. He will tell you that he "feels" it, but is unable to define what the sensation is. Likewise, one cannot explain some of the finer points in the art of lapping, and can only point out those which are fundamental, because proficiency and skill must be acquired by practice and experience.

54

Materials for Laps. — Laps are usually made of soft cast iron, copper, brass or lead. In general, the best material for laps to be used on very accurate work is soft, close-grained cast iron. If the grinding, prior to lapping, is of inferior quality, or an excessive allowance has been left for lapping, copper laps may be preferable. They can be charged more easily and cut more rapidly than cast iron, but do not produce as good a finish. Whatever material is used, the lap should be softer than the work, as, otherwise, the latter will become charged with the abrasive and cut the lap, the order of the operation being reversed. A common and inexpensive form of lap for holes is made of lead which is cast around a tapering steel arbor. The arbor usually has a groove or keyway extending lengthwise, into which the lead flows, thus forming a key that prevents the lap from turning. When the lap has worn slightly smaller than the hole and ceases to cut, the lead is expanded or stretched a little by the driving in of the arbor. When this expanding operation has been repeated two or three times, the lap usually must be trued or replaced with a new one, owing to distortion.

The tendency of lead laps to lose their form is an objectionable feature. They are, however, easily molded, inexpensive, and quickly charged with the cutting abrasive. A more elaborate form for holes is composed of a steel arbor and a split cast-iron or copper shell which is sometimes prevented from turning by a small dowel pin. The lap is split so that it can be expanded to accurately fit the hole being operated upon. For hardened work, some toolmakers prefer copper to either cast iron or lead. For holes varying from $\frac{1}{4}$ to $\frac{1}{2}$ inch in diameter, copper or brass is sometimes used; cast iron is used for holes larger than $\frac{1}{2}$ inch in diameter. The arbors for these laps should have a taper of about $\frac{1}{4}$ or $\frac{3}{8}$ inch per foot. The length of the lap should be somewhat greater than the length of the hole.

External laps are commonly made in the form of a ring, there being an outer ring or holder and an inner shell which forms the lap proper. This inner shell is made of cast iron, copper, brass or lead. Ordinarily the lap is split and screws are provided in the holder for adjustment. The length of an external lap should at least equal the diameter of the work, and might well be longer.

Laps for Internal and External Work. — The laps which are shown in Fig. 1 are, according to an experienced toolmaker, excellent designs for both the outside and inside lapping of cylindrical parts. At *A* is shown an inside lap with the arbor in place. The included angle of the taper of this arbor should be about 2 degrees; this is considered great enough for any kind of work. The lap proper, or the part that is in contact with the work, is made of bolt copper, and is shown in detail at *F*. Cast iron and lead are sometimes used, but copper will be found satisfactory

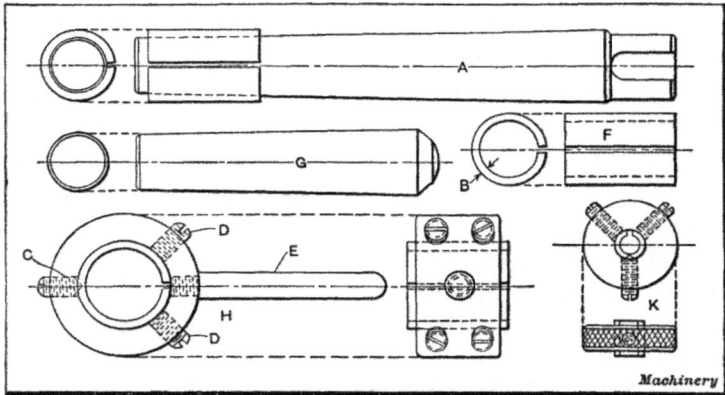

Fig. 1. Laps for Internal and External Work

for hardened work. The lap is split as shown, to allow it to expand as it becomes worn. The length of the lap should be somewhat greater than the length of the hole to be operated on, and the thickness *B* should not be more than $\frac{1}{6}$ or less than $\frac{1}{8}$ of the diameter of the work.

When making these laps, especially small ones, a hardened swedging plug (shown at *G*), ground to the same taper as the arbor, can be used to advantage for tapering the hole through the lap before it is turned and slotted. If in the operation of lapping, the hole becomes "bell mouthed," that is, enlarged at the ends, this is caused by the introduction of sharp emery from time to time as the hole is being lapped. To obviate this, the lap should be cleaned of all loose emery and expanded by driving the arbor farther into it. The hole is then dry lapped by using only the emery that sticks or is charged in the lap. This process must be repeated occasionally until the proper size is obtained. If the operator is careful to see that the emery used is not too coarse, and the lap is kept expanded to fit the work at all times, the result will be a straight hole.

At H is shown an outside lap. The proportions of the lap proper should be the same as were given for inside laps. The same method of procedure described for inside work should also be followed, viz., the lap must be freed from oil and loose emery from time to time as the work progresses. The pointed screw C keeps the lap from slipping out of place, and the adjusting screws D compress it to fit the work. A handle E should be used on all laps of large size, as it will be found much more convenient than a lathe dog, which some workmen use for moving the lap across the work. At K is illustrated an outside lap and holder for small work, say, less than $\frac{1}{2}$ inch in diameter. Laps of this size are not provided with a handle, but are knurled on the outside as shown.

It is the experience of many toolmakers that cast-iron laps are superior to those made of copper, and that lead laps are preferable to either under certain conditions. Some of the laps referred to in the foregoing, therefore, would, perhaps, not be practical in a shop where nearly every hole is finished by lapping, and by different classes of workmen. There are considerable variations in the practice of lapping and the best method often depends upon existing conditions. The material used for holding the abrasive is one of the most important factors when the work is soft; and it is very important that this material should be softer than the work. In many shops where nearly every hole is finished by lapping, lead has been adopted as the best metal to use. It is inexpensive and can be remolded by apprentices; besides it charges very quickly, which is a much desired and important feature. The lap arbor illustrated at A in Fig. 2, is used for holding the lead laps, which may be molded in as many sizes as desired. The molding arbor should be an exact duplicate of the working arbor. A small groove e is milled the entire

length of the molding arbor, producing a driving key *f* on the lap
which fits the working arbor.   This driving key is very necessary,
as it is almost impossible to prevent considerable friction between
the work and the lap.   Without the driving key, the lap would
surely revolve on the arbor and become tight in the work.

It is often very convenient to have laps with two or more diameters, as shown at *B*, which enables the operator to readily find a
size for each job.   It may be necessary, of course, to slightly

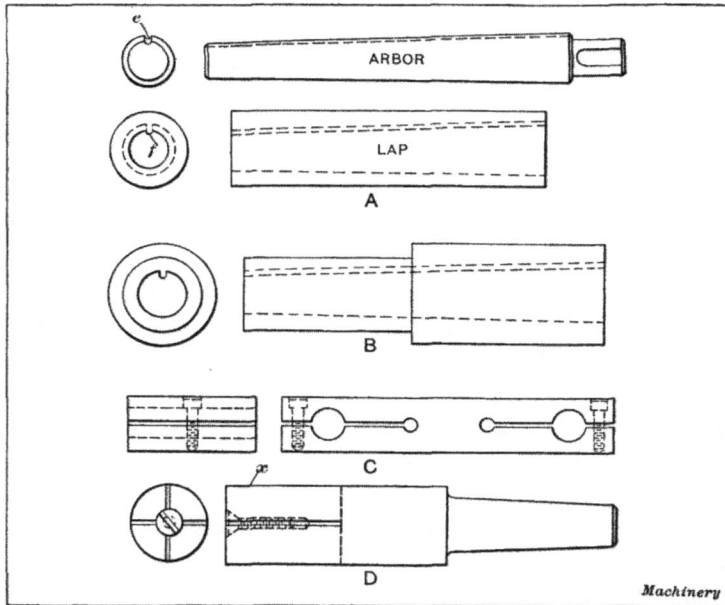

Fig. 2.   Lead Laps — Cast-iron Adjustable Laps

reduce one of the diameters to obtain the desired size.   The
usual custom is, when a lap is a little too small, to flatten it between two parallel plates in an arbor press, which forces the metal
outward on the open sides.

For outside lapping a cast-iron, adjustable lap similar to the
one shown at *C*, Fig. 2, will give good results.   Any piece of cast
iron will do, but it is better to have special castings of different
sizes, with holes in each end, as shown.   The slots and holes
serve to hold the loose particles of the abrasive.   The lower view

D shows the style of lap to use in a hole which bottoms, or does not extend through the piece to be lapped. It is slotted and tapped for an expansion screw, which makes adjustment very easy. The slots hold the loose particles of abrasive. This style of lap is usually made of cast iron, and if provided with a taper shank, it can be fitted to the drill press or lathe spindle. With this form of lap a hole which bottoms can be lapped straight, as the point of contact on the lap extends back from the end about $\frac{1}{4}$ inch to a point *x*; beyond this point it is slightly tapering.

Fig. 3. Laps and Lap-holders used for Making Plug Gages

Lapping Plug and Ring Gages. — For plug and ring gage work, cast iron is generally considered the best lap, and although it cannot be charged with abrasive as readily as copper or lead, it gives much better results, besides wearing much longer than the other metals. Laps for lapping plugs are made from disks ranging from $\frac{3}{16}$ to about $\frac{1}{2}$ inch in thickness, are drilled and reamed to a sliding fit on the ground plug, and split on one side to allow of adjustment. Several laps and lap-holders, such as are used for lapping plug gages, are shown in Fig. 3. These are similar to the external lap shown in Fig. 1. The holders, made in a few standard sizes to accommodate the different disks, are of machine steel, knurled, and have three adjusting screws to enable the operator to regulate the tension or size of the lap.

The piece to be lapped should be running at the speed required in grinding, which varies according to diameter, and the lap adjusted at all times to grip firmly on the surface, but sufficiently free to allow its being held by the fingers. In the case of large work a wood clamp may be used. As the piece revolves, the lap is slowly drawn back and forth from end to end, and under no circumstances should this oscillation cease while the plug is in motion.

The proper abrasive to use in this operation is flour of emery, or a very fine grade of carborundum; the latter, being the faster cutter, seems more desirable. It is mixed with sperm or lard oil, to the consistency of molasses, and applied sparingly to the surface being treated, from whence it is taken up by the lap, which becomes charged as it passes over after each application.

As the operation is almost completed, however, this is discontinued, and a drop or two of oil charged with the finest particles of flour emery is substituted. This is obtained by sifting about a tablespoonful of flour emery into a tumbler of lard oil, when, after standing an hour, the oil should be poured off, and will be found charged with the finest emery, the coarse particles having settled at the bottom. This abrasive is applied, a drop at a time, from the end of a small pointed stick or wire, and will make a remarkably smooth and bright finish.

Should it be necessary to remove an unusual amount of metal by the lapping process, much faster methods can be employed. For instance, lead or copper laps charged with a coarse abrasive liberally applied cut quite rapidly, but the results are not likely to be satisfactory if accuracy is desired.

Should there be any hitherto undiscovered soft spots in the work, they will invariably appear in the lapping, as their duller color is contrasted with the rest of the harder surface. However, it is possible for a piece to be slightly soft throughout, and finish up uniformly bright, but the softness should be discovered by file or other test at an earlier stage. In either case the piece can only be rehardened, and reworked for some smaller size, as a plug gage with a soft spot on its surface is useless.

Assume that a 1-inch plug gage is to be lapped to size. Such a gage needs only about 0.001 or 0.0015 inch for lapping. An outside lap similar to the type shown in Fig. 3 should be used. The flour emery, or other abrasive, should be sifted through a cloth bag to prevent any large particles of emery entering the lap and scratching the gage. After sifting the emery it is mixed with lard or sperm oil to the consistency of molasses, as previously stated. The gage is then gripped in the chuck of the lathe by the knurled end and a light coating of the abrasive is applied to the surface to be lapped. The lap is adjusted to fit snugly on the gage and the lathe is speeded up as fast as possible without causing the emery to leave the gage. The lap requires constant adjusting, to take up the wear of the lap, and reduction in size of the gage. This adjustment is effected by the screws in the holder. When measuring the gage, it should be measured at both ends and in the center to make sure that it is not being lapped tapering. When the gage has been lapped to within 0.0002 inch of the finished size, allow the gage to thoroughly cool and then by hand lap lengthwise of gage to the finished size. By so doing all minute ridges that are caused by circular lapping are removed, thereby leaving a true surface and also imparting a silvery finish. A gage should never be lapped to size while warm (heated by the friction of the lap), because the gage expands when heated, and if then lapped to size it will contract enough to spoil it.

The form of lap commonly used for ring gages is a cast-iron cylinder with a taper hole, split diagonally on one side to allow of expansion as it is forced on a taper arbor, to compensate for the gradual enlarging of the hole being lapped. A number of these internal laps and their arbors, are shown in Fig. 4. The lap should be about three times the length of the ring it is intended to be used in. The same rules regarding abrasive, speed, etc., apply as in the lapping of plug gages, but care should be exercised to avoid a too generous application of the abrasive as the process nears completion, for, if applied too lavishly, the particles have a tendency to crowd under the edges and cause a bell-mouth effect. This latter trouble is sometimes eliminated by making the rings with a small extension collar on each end which is ground off after the rough lapping has been completed;

but this is somewhat expensive, and, except for master rings, hardly necessary.

In the making of small ring gages which do not allow the insertion of a substantial cast-iron lap, a tool steel lap charged with diamond dust can be used. This abrasive, which is not extensively used outside the watch factories and concerns doing work of like nature, must be used to be appreciated. It can be purchased as Brazilian bort, in a pebbly form; it is crushed in a suitable mortar, and graded to suit requirements, and it is particularly applicable to hand or form laps, or laps for delicate or sharp corners. It is also a very rapid and smooth cutter, economical and lasting, and is readily taken up and retained on the surface of a tool-steel lap to which a very small quantity is

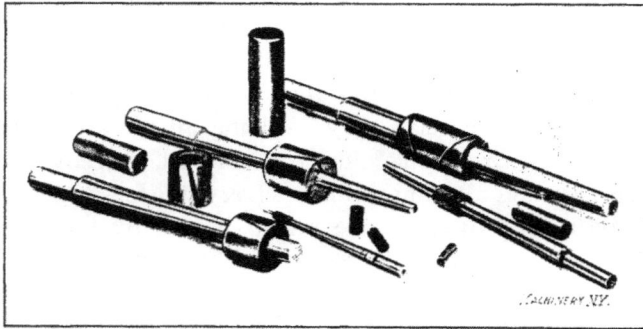

Fig. 4. Laps and Lap Arbors used in Making Ring Gages

applied mixed with sperm oil, and rolled in with an extremely hard roll. Occasional re-chargings are necessary as the work progresses, but in the intervals a drop of sperm oil is used on the lap.

In grinding out the inside of a ring gage, considerable difficulty is sometimes experienced in adjusting the grinder so that it will grind straight or cylindrical. One way to prove the straightness of a hole being ground is to move the wheel over to the opposite side of the hole until the wheel will just barely "spark." Then, beginning from the back of the hole, feed out, and if the hole is tapering, the wheel will either cease to spark, or will spark considerably more. Another way of testing the hole is to fasten a sensitive indicator to the spindle of the grinder and after placing the contact pin of the indicator on the opposite side of the hole, feed it in and out; the pointer will then record, in thousandths of an inch, just what the deflection is.

Lapping Conical Holes. — Taper or conical holes are sometimes lapped by using cast-iron plugs or laps having the same taper as the hole in the work. This method of finishing taper holes is liable to be unsatisfactory. In the first place it is difficult to secure a smooth surface, because the conical lap cannot be moved back and forth across the surface being lapped; moreover, the lap tends to cut annular grooves into the work, as it remains in one position, and imperfections in the hole will, to some extent, be transferred to the lap. When loose abrasive is used, a conical lap tends to produce a more abrupt taper in the hole being lapped, because the abrasive is gradually carried outward toward the mouth of the hole by the action of centrifugal force which, of course, increases as the diameter of the hole increases. Taper laps should be charged by rolling the abrasive into the surface, in the same way that cylindrical laps are charged. At least one roughing and one finishing lap should be used and if a smooth hole is necessary, several laps may be required. Slight errors in the taper are sometimes corrected by charging the roughing lap in accordance with the error; for instance, if the taper is slightly greater than it should be, the small half of the lap only is charged.

In making a ring gage having a taper hole or a taper plug gage, a different method of lapping should be employed. The facts regarding lapping are these: First, the lap should fit the hole at all times; second, the lap should constantly be moved back and forth. If a taper lap is made to fit the taper hole it will tend to stick fast and not revolve, and if held in one place, the lap will quickly assume the uneven surface of the hole. If the operator attempts to lap a taper hole by constantly revolving the gage on a straight lap, he will surely dwell longer in one place than another, thereby making a hole that is anything but round. The following method is, therefore, recommended: First grind the hole to size, plus the allowance for lapping; then, without disturbing the position of the slide rest or grinder head change the emery wheel for a lap made of copper (of the same shape as the emery wheel with the exception of having a wider face) and lap in the same manner as the hole was ground, care being taken not to "crowd" the lap by attempting to lap or grind too fast.

Laps for Flat Surfaces. — Laps for producing plane surfaces are made of cast iron. In order to secure accurate results, the lapping surface must be a true plane. Many toolmakers claim that a flat lap that is used for roughing or "blocking down" will cut better if the surface is scored by narrow grooves. These are usually located about $\frac{1}{2}$ inch apart and extend both lengthwise and crosswise, thus forming a series of squares similar to those on a checker-board. A rear view of a flat cast-iron lap is shown in Fig. 5. As will be seen, this particular lap is provided with ribs across the back like a small surface plate.

The first requisite of perfect lapping is a perfect lap, and right here is where the novice will make his first mistake; that is, in the preparation of the lap. To make a flat lap, it should be carefully planed, strains due to clamps being avoided, and then it should be carefully scraped to a standard surface plate. This is done by rubbing the face of the lap on the standard surface plate to obtain the bearing marks, and scraping down the high spots until a plane surface is obtained. If a standard surface plate is not at hand, a lap having a true plane surface can be obtained by planing three laps as accurately as possible and then scraping these three laps alternately, the same as when making surface plates. The procedure is as follows: Number the laps Nos. 1, 2 and 3; then, rub No. 1 and No. 2 together, and scrape the high spots until they fit. Then introduce No. 3 and scrape it down to fit No. 2, and then to fit No. 1, and so on. The third lap eliminates the error that might follow if only two laps were used. For example, it is possible to fit two plates accurately together without making them plane surfaces, one becoming concave and the other convex. The third lap absolutely prevents this and produces a perfect plane surface — if time and patience hold out. It is a slow operation, but not so slow as trying to lap a piece true with a lap that is not true. The laps may also be ground together instead of scraping. It is better to scrape them, because it is quicker than attempting to grind them level with the fine grade of emery that is required for nice lapping, and it must be remembered that when ground together the laps *are already charged;* hence, the necessity of using a fine grade of emery if they are ground together.

Method of Using a Flat Lap. — When lapping flat surfaces, No. 100 or 120 emery and lard oil (or some other abrasive of similar grade) may be used for charging the roughing lap. For

Fig. 5. Back of Standard Flat Lap, showing Ribbed Construction

finer work, a lap having an unscored surface is used, and the lap is charged with a finer abrasive. After a lap is charged, all loose abrasive should be washed off with gasoline, for fine work, and when lapping, the surface should be kept moist, perferably with kerosene. Gasoline will cause the lap to cut a little faster, but it evaporates so rapidly that the lap soon becomes dry and the surface caked and glossy in spots. When in this condition, a lap will not produce true work. The lap should be employed so as to utilize every available part of its surface. Gently push the work all around on its surface, and try not to make two consecutive trips over the same place on the lap.

Do not add a fresh supply of loose emery to a lap, as is frequently done, because the work will roll around on these small particles, which will keep it from good contact with the lap, causing inaccurate results. If a lap is thoroughly charged at the beginning, and is not crowded too hard and is kept well moistened, it will carry all the abrasive that is required for a long time. This is evident, upon reflection, for if a lap is completely charged to begin with, no more emery can be forced into it. The pressure on the work should only be sufficient to insure constant contact. The lap can be made to cut only so fast, and if excessive pressure is applied, it will become "stripped" in places, which means that the emery which was imbedded in the lap has become dislodged, thus making an uneven surface on the lap.

The causes of scratches are as follows: Loose emery on the lap; too much pressure on the work which dislodges the charged emery; and what is, perhaps, the greatest cause, poorly graded emery. To produce a surface having a high polish free from scratches, the lap should be charged with emery or other abrasive that is very fine. The so-called "wash flour emery," sold commercially, is generally too uneven in grade. It is advisable for those who have considerable high-class lapping to do to grade their own emery in the following manner: A quantity of flour emery is placed in a heavy cloth bag, and the bag gently tapped. The finest emery will work through first, and should be caught on a piece of paper. When sufficient emery is thus obtained it is placed in a dish of lard or sperm oil. The largest particles of emery will rapidly sink to the bottom, and in about one hour the oil should be poured into another dish, care being exercised that the sediment at the bottom of the dish is not disturbed. The oil is now allowed to stand for several hours, say over night, and then is decanted again, and so on, until the desired grade of abrasive is obtained.

Charging Laps. — To charge a flat cast-iron lap, spread a very thin coating of the prepared abrasive over the surface and press the small cutting particles into the lap with a hard steel block as indicated in Fig. 6. There should be as little rubbing as possible. When the entire surface is apparently charged, clean and examine for bright spots; if any are visible, continue charging until the entire surface has a uniform gray appearance. When the lap is once charged, it should be used without applying more abrasive until it ceases to cut. If a lap is over-charged and an excessive amount of abrasive is used, or more than can be imbedded into the surface of the lap, there is a rolling action between the work and lap which results in inaccuracy. The

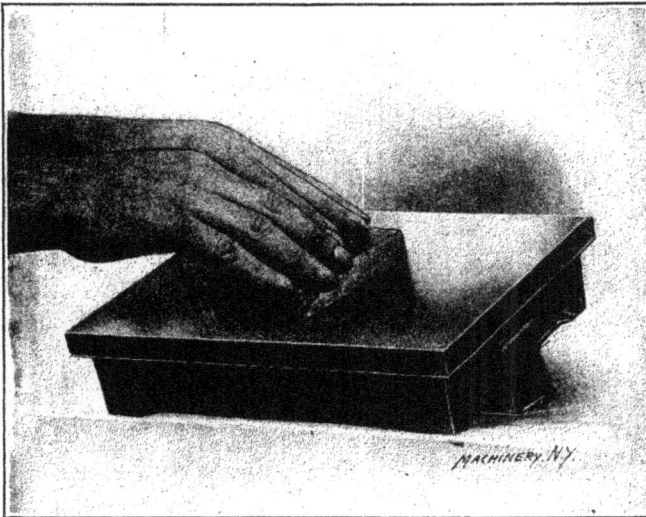

Fig. 6. Charging a Flat Lap, using a Hardened Steel Block

surface of a flat lap is usually finished true, prior to charging, by scraping and testing with a standard surface plate, or by the well-known method of scraping-in three plates together, in order to secure a plane surface, as previously explained. In any case, the bearing marks or spots should be uniform and close together. These spots can be blended by covering the plates evenly with a fine abrasive and rubbing them together. While the plates are being ground in, they should be carefully tested and any high spots which may form should be reduced by rubbing them down with a smaller block.

To charge cylindrical laps for internal work, spread a thin
coating of prepared abrasive over the surface of a hard steel
block, preferably by rubbing lightly with a cast-iron or copper
block; then insert an arbor through the lap and roll the latter
over the steel block, pressing it down firmly to imbed the abrasive into the surface of the lap.   For external cylindrical laps,
the inner surface can be charged by rolling-in the abrasive with
a hard steel roller that is somewhat smaller in diameter than the
lap.   The taper cast-iron blocks which are sometimes used for

Fig. 7.   Lapping the Jaws of a Snap Gage

lapping taper holes can also be charged by rolling-in the abrasive,
as previously described; there is usually one roughing and one
finishing lap, and when charging the former, it may be necessary
to vary the charge in accordance with any error which might
exist in the taper.

Lapping Gage Jaws. — A good method of lapping the jaws of a
snap gage is illustrated in Fig. 7.   The lap is made of cast iron
and is relieved as shown, leaving only a thin edge or flange on
each side to bear against the jaws.   The gage is clamped in the
vise of a milling machine and the circular lap is mounted on an
arbor inserted in the spindle.   As the machine table is traversed

back and forth, the lap passes over the entire surface of the jaw, grinding it down in the same manner as would be done with a cup emery wheel. Care must be taken to clamp the gage in the vise so as not to spring it.

A lap should be turned on the arbor on which it is to be used, for it is almost impossible to put a lap back on an arbor after it has been removed, and have it run true. Therefore, the lap should be recessed quite deeply, as shown, to allow for truing up each time the lap is placed on the arbor. Perhaps when the lap is mounted on an arbor in the milling machine, it will be found to run out not more than 0.001 inch, but that means that it is touching the work in only one spot, and the result can be hardly better than if a fly-cutter was used for a lap. When truing the sides of the lap a keen cutting tool is clamped in the vise and in this way the lap can be trued as nicely as though it were done in the lathe. In fact, it is superior, for there is absolutely no change in the alignment of the lap with the work spindle after it is turned, which might easily happen should it be turned in the lathe and then mounted in the milling machine spindle. With a perfectly true lap, a perfect contact between the lap and gage is insured for its entire circumference. Both sides of the lap should be turned at the same setting.

Figure 8 shows the operation of charging a circular lap, using a roller mounted in a suitable handle for the purpose. The emery is rolled in under moderate pressure. It is good practice to make the roller of hardened steel, and after charging the lap, all the surface emery should be thoroughly washed off.

The next step is to square up the jaws of the gage. Do not depend on the zero marks of the vise. The jaws of the gage may have sprung a little in hardening, and if the zero marks of the vise are depended upon to square the work, there possibly will not be sufficient stock on the jaws to clean up. Be very careful to set the gage by the surface of the jaws and to clamp it in the vise as previously mentioned, so that it is under no pressure tending to spring it out of shape.

When lapping avoid adding a fresh supply of abrasive to the lap, as it is not only injurious to the quality of the product, but it naturally increases the time required for lapping. To illustrate the action, suppose that an arbor is to be ground in a grinding machine, and that it is belted so that it runs with a wheel and at the same speed. Evidently no grinding action could take place, as there would be no difference in motion. The condition is very similar when loose emery is placed on a surface lap. The emery simply rolls around between the work and the wheel, and occasionally a piece of emery is imbedded in the lap long enough to scratch the work. While it may look as though

Fig. 8.　Charging the Lap with a Roller

the lap is cutting much faster, the truth is that it cuts slower and produces poor work.

In lapping jaws, some workmen rough-lap, and then finish by hand, but a better job will result when finished in the machine. It is poor practice to rough-lap a gage, using a coarse grade of emery, and then wash the lap and smear it with fine emery. Of course the lap is already charged with a grade of emery last used, and the act of putting on a supply of fine emery on the lap will not produce as good a surface as if the gage were finished without the fresh supply of emery, although the latter is of a finer grade.

Rotary Diamond Lap. — This style of lap is used for accurately finishing very small holes in hardened steel, which, because of their size, cannot be ground readily with an ordinary abrasive wheel. While the operation is referred to as lapping, it is, in reality, a grinding process, the lap being used the same as a grinding wheel. Laps employed for this work are made of mild steel, soft material being desirable because it can be charged readily. Charging is usually done by rolling the lap between two hardened steel plates. The diamond dust and a little oil is placed on the lower plate, and as the lap revolves, the diamond is forced into its surface. After charging, the lap should be washed in benzine. The rolling plates should also be cleaned before charging with dust of a finer grade. It is very important not to force the lap when in use, especially if it is a small size.

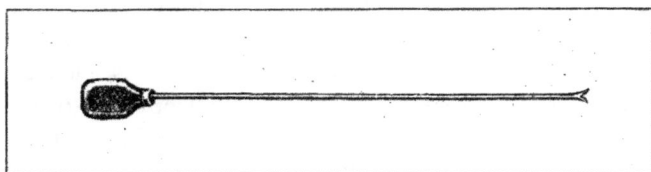

Fig. 9. Sound Magnifier

The lap should just make contact with the high spots and gradually grind them off. If a diamond lap is lubricated with kerosene, it will cut freer and faster. These small laps are run at very high speeds, the rate depending upon the lap diameter. Soft work should never be ground with diamond dust because the dust will leave the lap and charge the work.

When using a diamond lap, it should be remembered that such a lap will not produce sparks like a regular grinding wheel; hence, it is easy to crowd the lap and "strip" some of the diamond dust. To prevent this, a sound intensifier or "harker" should be used. This is placed against some stationary part of the grinder spindle, and indicates when the lap touches the work, the sound produced by the slightest contact being intensified.

One of these sound intensifiers is shown in Fig. 9. By using this simple instrument one can determine the instant when the lap touches the work. By placing the forked end on the work and the wooden part to the ear, the sound is greatly magnified, and it makes it much easier to determine the precise point of initial contact. If one depends upon the naked ear to tell when the lap touches the work, he is liable to crowd the lap too much, and scratch the work or strip the lap. With this instrument the mechanic will know the instant the lap just touches the work, and this is the position which it should occupy when in operation. The lap should not work under any appreciable pressure, but should simply touch the part being ground; hence the desirability of some means of magnifying the sound.

Grading Diamond Dust. — The grades of diamond dust used for charging laps are designated by numbers, the fineness of the dust increasing as the numbers increase. The diamond, after being crushed to powder in a mortar, is thoroughly mixed with high-grade olive oil. This mixture is allowed to stand five minutes and then the oil is poured into another receptacle. The coarse sediment which is left is removed and labeled No. 0, according to one system. The oil poured from No. 0 is again stirred and allowed to stand ten minutes, after which it is poured into another receptacle and the sediment remaining is labeled No. 1. This operation is repeated until practically all of the dust has been recovered from the oil, the time that the oil is allowed to stand being increased as shown by the following table, in order to obtain the smaller particles that require a longer time for precipitation:

To obtain No. 1 — 10 minutes.   To obtain No. 4 — 2 hours.
To obtain No. 2 — 30 minutes.   To obtain No. 5 — 10 hours.
To obtain No. 3 — 1 hour.   No. 6 — until oil is clear.

The No. 0 or coarse diamond which is obtained from the first settling is usually washed in benzine, and re-crushed unless very coarse dust is required. This No. 0 grade is sometimes known as "ungraded" dust. In some toolrooms the time for settling, in order to obtain the various grade numbers, is greater than that given in the table.

Making and Lapping Master Gages. — All measuring gages used in the factory of the South Bend Watch Co. are kept in repair and tested from time to time by means of standard thickness gages. These gages are made from $\frac{1}{2}$-inch diameter Sanderson drill rod, and vary in thickness from 0.025 up to and including 0.500 inch. Twenty-seven of these gages are shown wrung together in Fig. 10, this being possible owing to their wonderful accuracy. The way they were made and lapped is described in the following:

A piece of $\frac{1}{2}$-inch drill rod was held in a chuck and the gage blanks cut off, leaving sufficient material for facing, grinding and lapping. After hardening, the temper was drawn just enough to

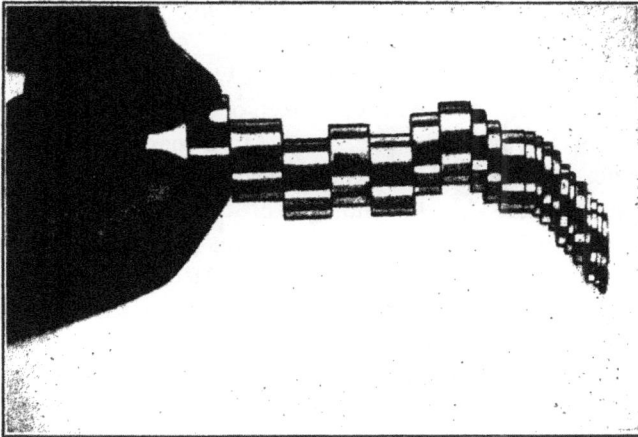

Fig. 10. Twenty-seven Standard Thickness Gages wrung together

remove the strains set up in hardening. Then the blocks were ground, leaving 0.0002 inch on a side to be removed by lapping. The cast-iron block to be used for lapping was planed as true as possible, using a flat tool for finishing; then, a hardened and ground steel block was used to rub diamond dust into the lapping block. Before describing the lapping of these gages it might be well to explain how the diamond dust was obtained. Splint and broken pieces of diamonds were put in a mortar, and crushed to powder by using a hardened steel pestle. The powder was then removed and placed in a receptacle partly filled with watch oil. The receptacle holding the diamond dust and oil was allowed to stand for ten minutes; then the oil was poured off and the sedi-

ment removed and labeled No. 1. The oil with the finer dust in
it was then allowed to remain in the receptacle thirty minutes,
after which the sediment was removed and marked No. 2. The
diamond dust still remaining in the oil was next allowed to settle
from 4 to 6 hours; then the dregs were removed and labeled No.
3. This process was again repeated and the oil allowed to stand
for 24 hours, the sediment obtained being No. 4. At this stage,
the diamond dust that had not sifted through the oil was very
fine, so that the receptacle was put away and left for two weeks.
The sediment was then removed and labeled No. 5. This is the
diamond dust which was used for lapping these fine thickness
gages.

The gages have a hole in them and were held in contact with
the lapping block by means of a piece of steel pointed like a lathe
center, and fitting in the hole, but not passing completely through
the gage. The gage to be lapped was held down firmly on the
diamond-charged block, and given a rotary motion. When
doing work of this kind, the pressure on the gage should not be
released before the lapping is stopped. After rotating the gage a
few times on the block, remove all foreign matter with benzine,
and repeat the lapping process until the gage is reduced to the
desired thickness. Here is a point about lapping with diamond
dust to be observed — never apply any more dust after the lap
has once been charged, and with benzine remove the material
ground from the work when the block begins to glaze. When a
lapping block begins to glaze, some toolmakers apply more dust,
but this is wrong. The reason for the glazing is that the material
removed fills the pores of the iron and prevents the diamond dust
from abrading the work.

These gages were lapped so accurately that they could be wrung
together and held without dropping apart, as shown in Fig. 10.
Of course, the more closely the surfaces approach true planes, the
greater will be the power required to separate them. This work
is not as difficult as it appears, but there is a slight "knack" in
lapping which can only be acquired by experience. However,
even a novice by following the foregoing method will be surprised
at the accurate results obtained.

Rotary Flat Lap. — A type of flat lap which has been used to some extent consists principally of a circular disk which forms the lap proper and revolves in a horizontal plane, a vertical driving shaft at the top of which the circular lap is attached, suitable means for driving the vertical shaft, and a stand or frame for supporting the parts mentioned. Good results have been obtained with a rotary lap having a lap proper made of cast iron, and faced with lead on the lapping surface. The disk was provided with anchor grooves and the lead, after having been poured onto the surface, was hammered to make it more compact. The disk should be surrounded by a sheet-metal guard to prevent the oil and abrasive from flying about. The vertical shaft may be driven direct by a quarter-turn belt or through bevel gearing from a horizontal shaft, driven in any convenient way. It is important to so use the surface of the rotary lap that it will be kept straight and true. The outer edge runs so much faster than the inside that hollow places will be worn in the surface unless care is taken to prevent this; it is a good plan to test the surface frequently with a straightedge and then distribute the lapping so as to reduce whatever irregularity may exist. An attachment for a rotary lap which is useful for lapping the ends of round or square bars, etc., consists of a horizontal guide-bar which extends across the center of the lap and is located a slight distance above the lapping surface. This bar has a small sliding head which can be moved from the center of the lap out to the edge, and in this sliding head there should be a V-shaped groove located at right angles to the surface of the lap. When lapping the end of a rod or bar the work is supported in the groove of the sliding head and at right angles to the surface of the lap. A rotary lap may be charged by sprinkling the abrasive over the surface when the lap is not in motion and then pressing in the abrasive by rubbing it with a piece of round iron held in the hands. An old pepper box with a perforated top is excellent for sprinkling the abrasive over the lap surface. As to speed, it was found that a 24-inch rotary lap should revolve at about 300 revolutions per minute.

CHAPTER III

MAKING STRAIGHT AND CIRCULAR FORMING TOOLS

Forming tools are made in either straight or circular shapes. Forming tools for the lathe or planer are ordinarily made flat or straight. In some cases a flat formed blade is bolted to a holder, but tools that are to be used very little are often made solid, the formed cutting edge being machined and filed on the flat forged end of the tool. When a number of different tools are needed, it is more economical to make one shank or holder and attach separate cutters or blades of the required form.

The diagram A, Fig. 1, shows a straight forming tool of the vertical or straight-faced type. This style of tool is used on automatic turning machines, etc., especially for large work. As the plan view shows, the rear side is dovetailed to fit the holder which, in turn, is attached to the cross-slide of the machine. The circular type of forming tool B is used in preference to the flat or straight type for many classes of work, especially in connection with automatic screw machine practice, because it is easily duplicated after a master tool for turning it is made. The circular tool can be ground repeatedly without changing its shape. The straight form of tool may also be ground repeatedly without affecting the shape, when it is made with a formed surface which is of uniform cross-section; some straight or flat tools, however, especially when made for producing a comparatively small amount of work, are filed to shape before hardening and when the top of the tool is ground down for sharpening, the shape of the cutting edge changes owing to the clearances. A straight-faced tool A can be made without clearance on angular or curved surfaces, but edges which are at right angles with the axis of the work should preferably be given a slight side clearance; this clearance angle, however, will affect the accuracy of the tool very little as the top is ground away for sharpening, and the error resulting from this cause could be disregarded on many classes of work.

Forming tools are also made which operate tangentially instead of radially; that is, the cutting edge, instead of moving in toward the center of the work, moves along a line tangent to the outside surface being formed. The tool may be mounted on a vertical slide operated by a hand lever, or it may be fed horizontally. When using the cutters shown in Fig. 1, the entire cutting edge comes into action and, in some cases, this results in chattering and springing of the work. With a tangent cutter, it is

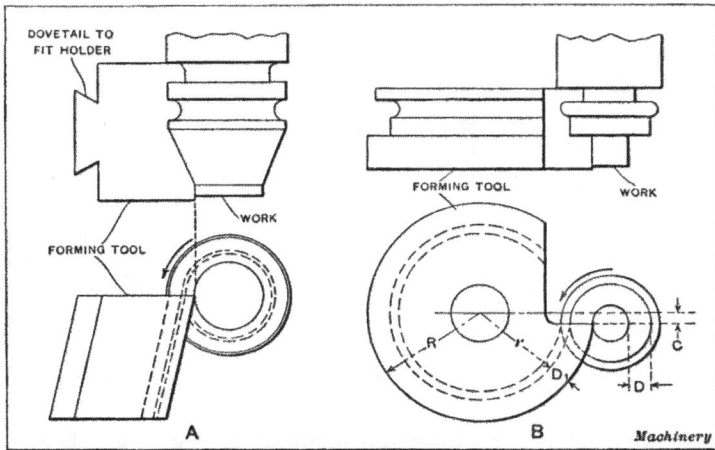

Fig. 1. (A) Straight Forming Tool of Vertical Type. (B) Circular Forming Tool

possible to prevent this trouble by beveling the front end so that the point of the tool begins cutting first and then passes beyond the center of the work, while the remaining part of the inclined cutting edge gradually comes into the working position.

There are many different methods of making forming tools, but little has been written on this subject in a way which will enable the inexperienced toolmaker or machinist to compute the distances, diameters, or angles so as to produce the required dimensions on the finished work. For instance, if a circular tool is to produce certain diameters on the work, and we transfer the exact ratio of these different diameters from the drawing to the forming tool, we will not be able to get the required dimensions when the cutting edge of the tool is one-quarter of an inch below the center. If we have to make a straight forming tool which, in the machine, is to stand at an angle of 15 degrees, we can, when it is not very wide, use a master forming tool. When, however, the tool is very wide, so that the use of the master tool is impracticable, and it is necessary to mill or plane it to shape, some computation is then necessary in order to make the shape such that it will produce the required dimensions on the work when the tool is held at an angle in the machine. The calculations con-

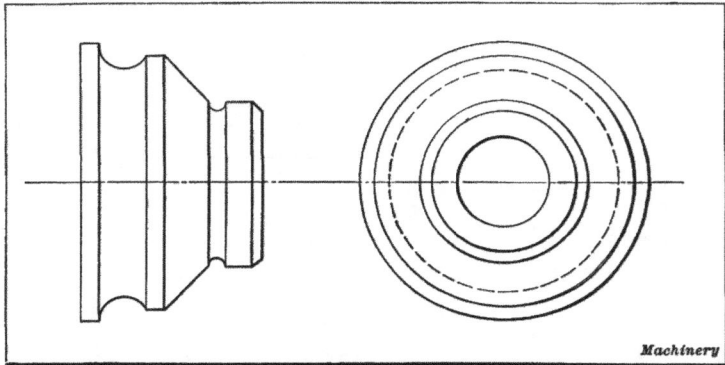

Fig. 2. Example of Forming Tool Work

nected with straight and circular forming tools will be referred to later.

Making Straight Forming Tools. — The following method of making straight forming tools has proved satisfactory and will produce accurate results if the work is done carefully. Suppose that we are called upon to make the tools for producing with accuracy such a piece of work as shown in Fig. 2, and that a master former is to be made so that at any future time the forming tool can be duplicated at small expense.

The master former may be made as in Fig. 3, being an accurate duplicate of the model with the addition of a shank about an inch and a half in length on each end, these shanks serving as an arbor for the tool. The formed part is milled down to the center, to produce a cutting face for future operations, after which the tool is hardened, tempered and the face accurately ground. To facilitate this face grinding the fixture shown in Fig. 4 is employed, the work being done on a surface grinder. The most essential point to be observed in grinding such a former is to have the cutting face radial, and this is accomplished by the use of this fixture. The fixture is placed on the grinder so that the line of centers is at right angles with the grinding wheel; the center c is removed from the block d, and the grinding wheel is brought to bear on the ball b, which is a running fit in the lever a. This lever is fulcrumed on the block d and is held upward by means of a spring attached to the short arm. With the grinding wheel at rest, the

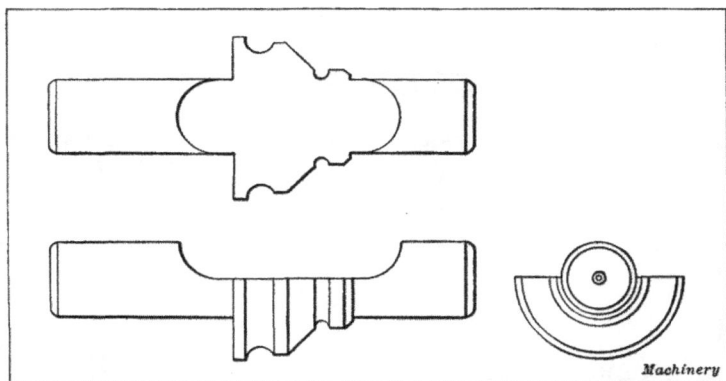

Fig. 3. Master Former used in Making a Straight Forming Tool

table of the grinder is now run back and forth and the wheel fed downward until it reaches such a position that when it passes over the ball b, the front end of the lever will indicate zero. This shows the operator that the periphery of the wheel is in perfect alignment with the center of the fixture. The center c is then replaced and the master former, with dog attached, is placed between the centers and ground. By the use of the handle e, which engages the knurled head of the adjustable center, the former is turned slightly after each cut across the face until a keen cutting edge on the former is obtained.

The block from which the forming tool is to be made is placed in the vise of the milling machine and roughed out as near to the formed shape as possible, after which the master former is substituted for the milling cutter, as shown in Fig. 5. The cutting face of the master former must stand at the same angle with the vertical as the forming tool is to stand when placed in the screw

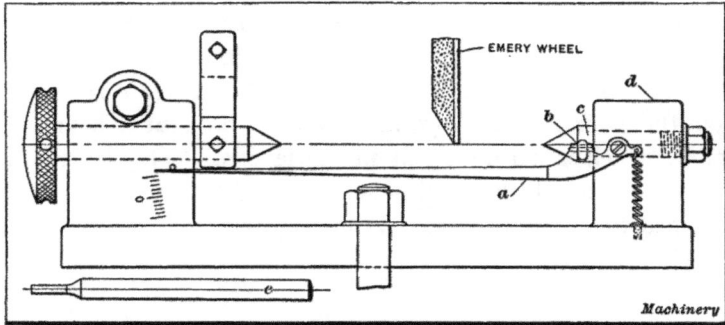

Fig. 4.   Device for Grinding Master Former

machine, and it is very essential to observe this point if accurate results are desired. When the proper angle of the former has been obtained, it is secured in position by a wooden wedge tapped

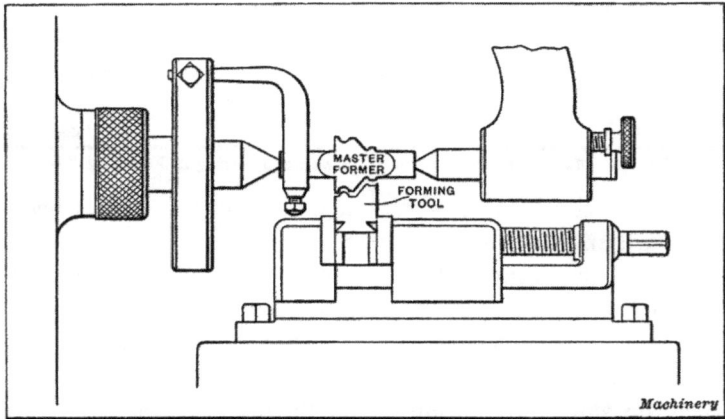

Fig. 5.   Making Straight Forming Tool from Master Former

in between the cone and the frame of the milling machine. The table is then run back and forth and the forming tool gradually cut to the desired shape by taking light scraping cuts with the master tool.

Straight forming tools are sometimes machined in the shaper.

One method consists in filing a templet to fit the model and from this making the shaping tool, which, in turn, is placed in the shaper with the face standing at the same angle as will the forming tool when placed in the screw machine. Then the forming tool blank is put in the vise and shaped in the usual way. As the shaper tool inclines backward, the cutting edge has "negative rake" and works with a scraping action so that the cuts must be very light. Before using the shaper forming tool, it is usually advisable to rough out the blank to be formed to approximately the required shape, so that there will be little metal to remove by the forming tool.

The foregoing methods are based upon the duplication of the formers by mechanical means, but when making straight forming tools by the method described in the following, the dimensions of the tool are computed. We have already made it clear that an error will exist if we transfer to the tool the exact differences in the various radii on the work, and it is to overcome this error that we subtract from the dimensions of the work such differences as are caused by the tool standing at an angle in the machine, as will be explained more fully later.

Making Vertical Forming Tools. — The methods referred to in the following represent the practice at the Windsor Machine Co., in the production of the vertical forming tools such as are used on the Gridley automatic machines. These forming tools were made at first by being shaped to the form required, it being necessary for the mechanic to use his best judgment in allowing for changes in hardening, but as tools of different shapes were not affected alike by these changes, it was impossible to be sure of accurate results. To overcome this defect, the vertical style of forming tools are ground after the hardening operation. This grinding of the hardened tools has practically solved the problem of accuracy and has made it possible to turn parts of much more intricate shape, and secure true corners, curves, and angles on the work.

One of the most frequent questions arising in regard to grinding forming tools is the expense, many believing that it takes more time than is actually required and that it is not advisable to

grind the tools owing to the cost.   Within reasonable limits, the
time should not be considered, because when a vertical tool is
completed, it will produce hundreds of parts, to say nothing of
the accuracy of production and duplicating qualities of the work.
At first thought, one might charge all the time consumed in grinding plus the time it would take to finish a tool in the old way, but

Fig. 6.   Grinding a Hardened Forming Tool on Surface Grinder

this is incorrect as much time is saved on the shaper work because it is not necessary to plane the form so accurately when the
tool is to be ground afterward, and coarser feeds may be employed.   Because of this advantage, very little extra time is
required for grinding, and this additional time is certainly offset
by the accurate surfaces and measurements obtained on the work.

Before grinding a forming tool, it is first roughed out on a shaper, 0.010 or 0.015 inch being left for grinding. The back of the tool or dovetail is finished to size so that no unnecessary grinding is required. The grinding is all done on a surface grinder fitted with a magnetic plate or chuck. The fixture for holding the tools while being ground is a very simple device and is illustrated in Fig. 6. It is essential that the top and bottom of the fixture be perfectly parallel and also that the fixed edge of the dovetail be true with both edges or sides of the base. If the straightedge on the magnetic chuck is true, obviously this holder can be removed and replaced without affecting the accuracy of the work; it can also be changed end for end, although care should be taken that the holder and magnetic chuck are clean when such changes are made. By inserting a thin piece of steel of the required thickness between the straightedge and fixture, the necessary clearance for the vertical faces of the tool can also be obtained, as will be described more fully later. A few of the many shapes of vertical forming tools that have been ground at the Windsor Machine Co., are shown in Fig. 7. In order to illustrate the general method of procedure in producing these forming tools, the different operations on the particular tool shown at *G* will be described in detail, as this tool is typical in many respects.

Dimensioning and Laying Out a Forming Tool. — When making the drawing of a forming tool, the draftsman should make each horizontal measurement correspond exactly with each relative measurement of length on the article to be turned. The vertical measurements should, preferably, be given at right angles to the front face of the tool, so that the mechanic can work directly from the drawing; in this way, the liability of mistakes is lessened and the work can be done more rapidly. Of course, these dimensions at right angles to the face of the tool vary slightly from the actual dimensions on the work, owing to the clearance angle, and must be calculated, as explained in the paragraph, "Vertical Forming Tool Calculations."

The exact method of procedure when laying out a tool depends, of course, largely upon the form of the cutting edge. Fig. 8

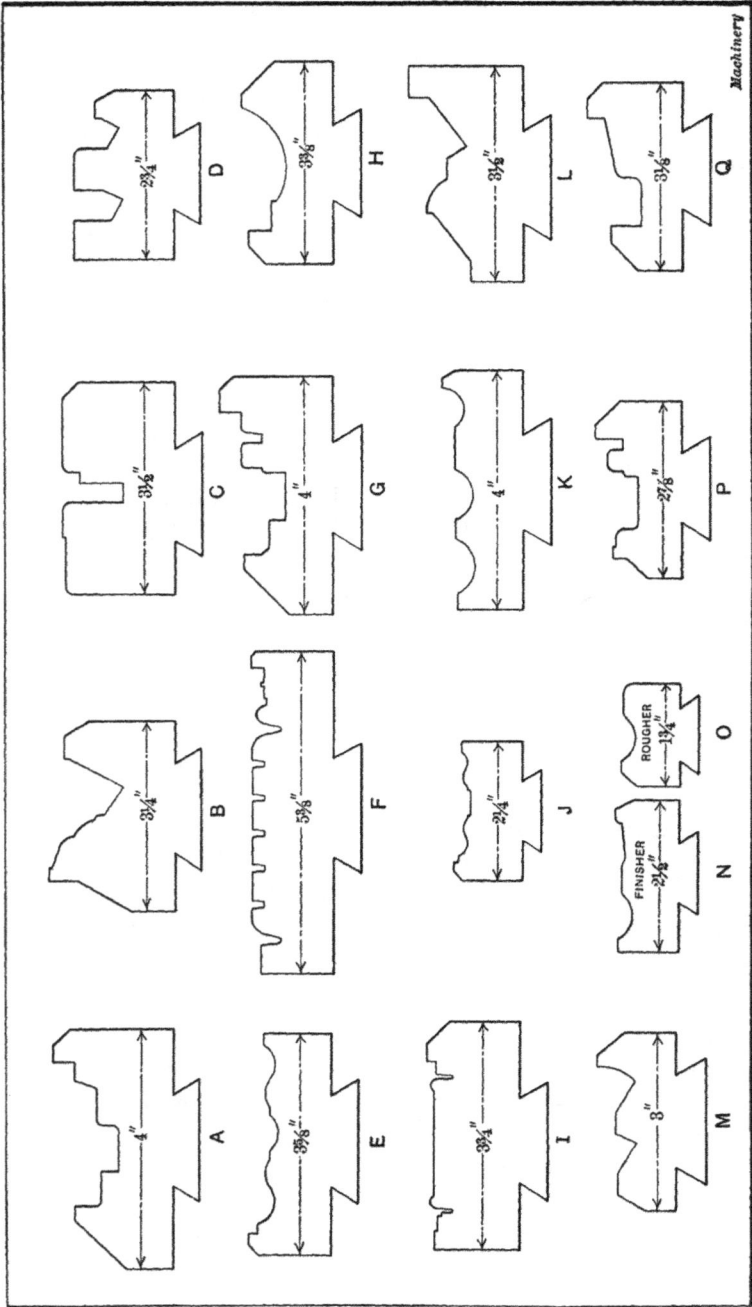

Fig. 7. Various Shapes of Vertical Forming Tools which have been Finished by Grinding

illustrates, in a general way, however, how this work is done. The tool shown by the diagrams A to F, inclusive, is similar to the one illustrated at G in Fig. 7. After copper-plating the front end or working face of the tool (by using a copper-sulphate solution) place the block on a surface plate with side x down; then with a scratch block resting upon a combination of size blocks equal in height to dimension a (see sketch B) plus 0.010 inch, scribe the first of the vertical lines (sketch J indicates how the scratch block is used in conjunction with the size blocks). By using different combinations of size blocks, scribe each succeeding vertical line until all the lines are laid out as indicated at B. The first measurement a is the only one that is made 0.010 inch greater than the dimension on the drawing, this allowance being left for grinding the left side; the distances between the other lines must be precise. Next place the block on surface y (see sketch A), and scribe, in the same manner, all horizontal lines, as indicated at C; then with a protractor set at the required angle, draw lines c–d and e–f, and also the two corner angles, as at D. With dividers, scribe arcs of the correct radius in each corner that is to be rounded. The block is now ready to be roughed out on the shaper.

It is needless to dwell on the method of removing the stock in the shaper, but when performing this operation, allowances for grinding should be made (as indicated by the dotted lines on sketch E, Fig. 8), 0.007 or 0.008 inch being left on a side; that is, measurements between faces such as g or h (sketch F) should be left about 0.015 inch narrow, while a measurement like k should be about 0.015 inch wide. No allowance need be made on vertical measurements because the same amount can be ground from each surface, so that there will be no change in the measurements. When roughing the tool out in the shaper, it is preferable to plane the surfaces straight or parallel, making no allowance for clearance on the side faces. The amount to be removed for clearance is so slight that it can easily be ground. When the tool has been roughed out, the upper end or working face should be beveled off to the required clearance angle which, at the Windsor Machine Co., is made 12 degrees.

Set of Wheels for Grinding Forming Tools. — When grinding forming tools, the general rule for the selection of wheels, *viz.*, use soft wheels for hard metal and *vice versa*, cannot always be strictly adhered to. Numerous experiments have shown that excellent results in grinding are obtained by using alundum wheels, different grades and grains being selected according to requirements. The following list gives an ample variety and all the shapes that are required may be obtained by using one wheel or the other and dressing the wheel face off to suit the shape desired on the tool.

Style.	Diameter.	Width.	Grain.	Grade.
Straight.............	7	½	46	H
Straight.............	6	½	46	K
Straight.............	6	1	46	H
Straight.............	6	¼	46	K
Straight.............	6	½	46	M
Cup.................	6	1	46	I
Cup.................	4	1⅜	46	K
Cup.................	4	1⅜	60	K
Saucer.............	6	9⁄16	46	K
Saucer.............	6	9⁄16	60	K

Quite a number of wheels is necessary to secure economical results, but when a stock of wheels is secured the consumption is small if they are handled properly. When first starting to grind form tools, one should not make too radical a change in the shape of a wheel face, but secure another size which is more suitable. In this way, there will soon be a variety large enough so that a wheel can be found which will require but little dressing to reproduce almost any form desired.

Method of Grinding a Forming Tool. — After the tool is hardened, it is ground to the finished dimensions. The first operation is to grind the upper end or working face of the tool. Enough should be ground off to insure sharp corners or edges, without having to grind too much on the formed surfaces. The next operation is that of grinding the formed part of the tool. The dovetailed clamping surfaces should first be inspected to see if they are clean. The tool is then clamped in a fixture similar to that shown on the surface grinder in Fig. 6. The tool shown in this illustration, however, does not correspond in form with the one selected as an example and illustrated in Fig. 8. Naturally, the first surface that should be ground is the one that is highest, thus providing a resting place for the base of the depth gage. All vertical measurements (as held for grinding) should be given from this surface. For measuring vertically when grinding, a micrometer depth gage has proved satisfactory, while for horizontal measurements, a set of size blocks may be employed.

If there is a dimension such as h of, say, $\frac{3}{4}$ inch (see sketch F, Fig. 8), a 7- by $\frac{1}{2}$-inch wheel of 46-H grade and grain would be about right for grinding. First grind the lower horizontal surface to within 0.001 inch of the required depth; then insert a thin strip of steel between the work-holder and straightedge of the chuck, in order to secure the necessary clearance for the sides of the tool. Use the sides of the wheel to rough off the sides of slot h to within 0.003 or 0.004 inch of the width at the narrowest end (working face) of the tool, grinding first the left side and then the right side. Next, raise the wheel and grind off the top surface of k and also surface N, allowing 0.001 inch for finishing. Adjust the work laterally and rough off surface O; then drop to surface M and grind that to 0.001 inch less than the required depth. Next, raise the wheel far enough to clear the beveled corner to the left of M, and rough off the side surface, care being taken not to grind too much in order to retain dimension g at the narrow end, in proper relation to slot h. Be sure that the strip of steel is at the proper end of the fixture to allow a clearance on this side.

Now take a 6- by $\frac{5}{16}$-inch saucer wheel (60-K) and dress off the face and sides, carefully retaining a sharp corner. With the fixture held as for the previous operation, grind the left side indicated by dimension h, great care being taken to grind it smooth and straight. While grinding, the lower edge of the wheel should be down even with the lower surface, and when the side is smooth, feed the wheel down far enough to take off the 0.001 inch from the lower surface that was left when rough-grinding; then feed the wheel to the right, thus finishing the lower surface as far as possible without touching the right-hand side of the slot. Note the reading of the handwheel dial and raise the grinding wheel

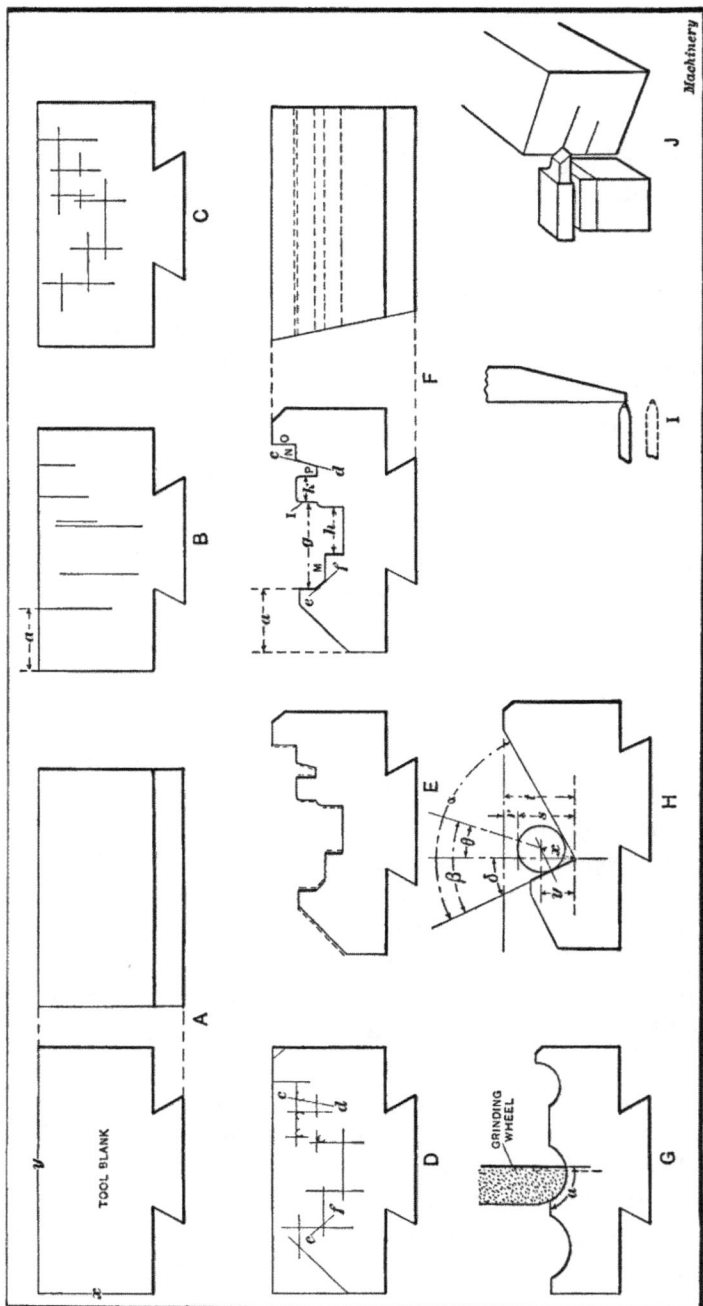

88   Fig. 8.   Diagrams Illustrating Method of Laying Out a Forming Tool, Grinding Curved Surfaces, and Gaging Angular Tools

one or two thousandths inch. Turn the fixture around, adjust the strip for clearance and then grind the other side of slot h until the required width is obtained. Of course, all measurements should be made at the working end or top face of the tool. Continue the grinding operation by feeding down to the same reading previously obtained and then feed horizontally to join the lower surface that was ground before turning the fixture around, thus finishing the slot. Raise the wheel and grind surface I and then side O, measuring from the slot h. Feed the wheel down far enough to take off the 0.001 inch left when roughing; then feed across surface N and the top surface of the part k.

Turn the fixture around again and adjust the strip for clearance. Place a 6- by $\frac{3}{8}$-inch 46-K wheel on the machine and bevel off the left corner of the wheel so that the width of the chamfer is a little less than the width of chamfer e–f on the work. Dish the side of the wheel slightly, so that only the corner does the cutting, and grind the remaining side of g until the width is correct; then feed down until the upper corner of the chamfer is ground to the right depth, after which move the wheel a little to the right and feed down, grinding horizontal surface M to the required dimension; then feed to the left until this horizontal surface joins with the beveled corner. Next dress off the back of a 60-K saucer wheel to the proper angle for side c–d, using the fixture for angular dressing illustrated in Fig. 9. (This fixture will be described later.) Make the face of the wheel slightly narrower than the width of slot P at the bottom; then with work-holding fixture inclined horizontally to give clearance to the side of the slot, grind the left side of slot P until k is of the required width, feeding down until slot P is of the proper depth. Raise the wheel slightly, remove the clearance strip and with the fixture set parallel to the travel of the table, grind the angular side, feeding laterally and down until slot P is of the correct width when the index reading is the same as when finishing the other side. Care must be taken not to touch the straight side when grinding the angular surface, because the cutting edge would be the first to be ground away by the wheel, thus spoiling the tool. This slot can be roughed out with a coarser wheel, if desired, but should be finished with a 60-K to retain the sharp corners at the bottom. In this case, it is not likely that a roughing wheel would be economical to use, because changing wheels and forming or dressing two wheels would off-set the extra time required for grinding with a 60-K wheel. Secure a 46-K saucer wheel and form it for grinding the small round corners. The work-holding fixture should be set in

Fig. 9. Fixture on Magnetic Chuck of Surface Grinder is used for Dressing Grinding Wheels at an Angle

different positions so that the same wheel may be used on all the corners of the same radius. The tool is now finished, except-ing the beveled surface at the left, and the left side. The beveled surface should extend a little below the largest diameter of the part to be turned, or, in this case, a little below the bottom surface of slot *h*.

In measuring a slot with the size blocks, insert the block or combination of blocks in the slot near the widest end and move them toward the working end of the tool as far as possible. Continue grinding from the side until the block will just pass through at the cutting edge.

When forming tools were first ground at the Windsor Machine Co., a holder that swiveled was employed, but it soon became evident that this holder was unnecessary and, in fact, undesirable owing to the time taken to change it back and forth, to say nothing of the possibility of a mistake caused by not getting it true for different surfaces. It is much preferable to have a plain holder like the one illustrated in Fig. 6 and then use a fixture to dress off the emery wheel to the required angle.

Grinding Angular Surfaces. — The fixture used for dressing grinding wheels at an angle (illustrated in Fig. 9) consists of a block which holds the diamond and is traversed back and forth on an inclined block; the latter, in turn, swivels on a stud at the center of the fixture. The outer end of the inclined block rests on an adjustable screw, which, being movable in a slot, allows adjustment for any position from the vertical to the horizontal. There are two of these adjusting screws so that settings for two angles can be made at the same time, one on either side of the vertical. This swiveling block is set at any required angle relative to the top of the fixture, the top being ground parallel to the base. The fixture is about 4 inches wide, thus giving a good bearing surface for the bevel protractor. A vernier protractor can be set practically to within $2\frac{1}{2}$ minutes of the given angle, and very few form tools require greater accuracy, although, for very fine work, the sine bar may be used in setting the swivel block. By using this truing fixture, one can dress a surface $\frac{1}{32}$ inch wide just as accurately to a given angle as a surface 4 inches wide.

The grain and grade of wheels used for grinding surfaces on an angle are governed by the same condition as for straight grinding, except for a forming tool having two angular surfaces which, at the point of intersection, form a sharp corner as, for example, a tool for turning bevel gear blanks. For such work it is well to use a 46-M wheel; this wheel may seem too hard at first, but experience has shown that the best results are obtained with this grade. As little as possible should be left for grinding, and when using a wheel of the grade mentioned, rough out to within 0.001 or 0.002 inch of the size; then, after re-dressing the wheel, the surfaces may be finished smooth and a sharp corner obtained.

The best way to grind forming tools which are to be used for turning bevel gear blanks and similar shapes, is to take a 46-M wheel, say, $\frac{1}{2}$ inch wide, and dress it down so as to make the width of each bevel in proportion to the width of the surface each side is to grind. The bottom of the angular groove is ground first by setting the wheel central, as indicated at A, Fig. 10, then

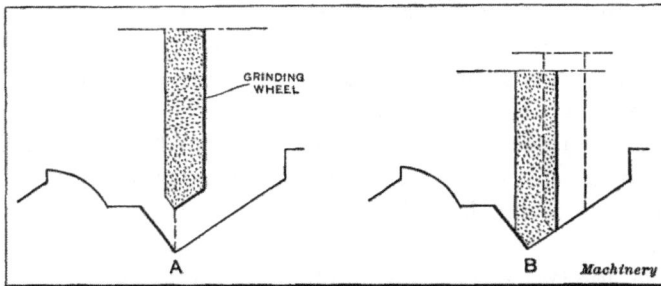

Fig. 10.   Method of Grinding an Angular Surface

feeding the wheel down and cutting to the full width of both bevels until the point of the wheel is about 0.002 inch less than the required depth. To grind the remaining surfaces, raise the wheel and move it sideways so that when it is fed down again, the wheel face will slightly overlap the surface already ground, as indicated by the dotted line in sketch B. The wheel is then moved farther up the incline, and so on until the entire surface is finished. In a similar manner, the opposite angular surface is ground. For finishing, re-dress the wheel and repeat the foregoing operation, taking off the 0.002 inch left when roughing. If these operations are performed carefully, a tool can be ground quite quickly without burning it, although it will become quite warm. The angular form of tool and those having large or irregular curves are the only ones that need a templet to work to, as all others can be gaged by direct measurement, using the graduated dials of the grinding machine and regular measuring tools.

When grinding sharp square corners, dress the face of the wheel off with a diamond attached to the magnetic chuck. When dressing the side of a wheel, the diamond should not be carried beyond the periphery of the wheel as indicated by the dotted lines in sketch I, Fig. 8, but should stop just before reaching the edge or in the position shown by the full lines, because the diamond breaks the corner of the wheel more or less, if it passes beyond the periphery; therefore, it should be brought up close to the edge and then be returned. A little practice will enable one to grind sharp corners.

Gaging an Angular Tool. — When an angular tool for bevel gear blanks does not have a curved surface in conjunction with the angular surfaces, a mechanic versed in trigonometry can save time by a few simple calculations. The diagram H, Fig. 8, illustrates how the depth of a V-groove can be accurately gaged by using a cylindrical rod and calculating the distance r to the top surface from which the measurement is taken. For a tool of the form illustrated, the two angles and the distance t are always given. As depth t is down to a sharp corner, it is difficult to determine it by direct measurement but by laying a cylindrical rod in the groove and measuring down to this rod, an accurate measurement may be obtained. As dimension r is the one required, the distance s from the top of the rod to the apex of the angle must be determined. As a rod lying in a V-slot will always center itself, a line drawn through the center of the rod and the apex of the angle will be equidistant from each side of the angle, thus giving the angle β which is one-half of angle α. A line drawn through the center of the rod and through the point of contact on either side of the angle, will be at right angles to that side. This line, and the one extending from this point of contact to the apex of the angle and the line extending from the apex to the center of the rod, forms a right-angle triangle of which one side (radius of the rod), and angle β are known, so that dimension x may be obtained. (Hypotenuse

equals side opposite divided by sine.) Drawing a vertical line
through the apex of angle α and also one at right angles to this
line through the center of the rod, gives a right-angle triangle of
which dimension y is desired. Angles β and δ being known,
angle θ is obtained easily, and, having dimension x, dimension
y may also be obtained. (Side adjacent equals hypotenuse
times cosine.) Therefore s will equal y plus one-half the diameter
of the rod, and this dimension, subtracted from t, will give the
required measurement r.

Points on Grinding Vertical Surfaces. — When grinding the
sides of vertical surfaces of a forming tool, any neglect or care-
lessness will be manifest in the action of the tool when it is being
used. Care must be taken to see that the wheel is not allowed
any side play when finishing vertical surfaces, as otherwise
when the wheel leaves the tool at the end, it will grind the
corners back on account of the gradual lessening of the wheel
area in contact, thus forming a convex surface adjacent to the
cutting edge. This will allow the stock, when the tool is in
use, to rub on the side below the cutting edge, which causes a
"loading" of the tool and makes it impossible to secure a good
finish on the work. As this lack of clearance caused considerable
trouble, the Windsor Machine Co. adopted the plan of grinding
a clearance on all straight side surfaces, to avoid any rubbing of
the work or loading of the tool when in use. This clearance is
obtained, as previously mentioned, by inserting a thin strip of
steel between the edge of the grinding fixture (see Fig. 6), and
the straightedge on the magnetic chuck, at whichever end of
the fixture needs to be inclined outward in order to provide
clearance for the cutting edge. This strip should be thick
enough to provide a taper of about 0.035 inch per foot, which is
approximately the minimum amount for good results. While
more clearance would be better as far as the cutting action of
the tool is concerned, it would result in too great a change of
length as the tool is ground away on the top for sharpening.
In most cases, however, a difference in length of 0.003 to 0.005
inch one way or the other is immaterial, especially on large
work. Obviously, this liner can be inserted at either end of the

holder, depending upon which side of the tool is being ground for clearance. This clearance is only necessary on straight sides as all curves and surfaces on an angle, no matter how slight, will clear themselves. The only exception as to curves is when one joins a straight side; then it should be ground at the same time that the straight side is ground for clearance.

Joining a Vertical Side and Beveled Surface. — Another grinding problem met with when finishing a forming tool similar to that shown at *L* in Fig. 7, is that of joining a straight or vertical side and a bevel, grinding the clearance on the vertical side and, at the same time, neatly joining it with the beveled surface. First, all the surfaces of the forming tool should be ground straight or without clearance; then a thin strip of steel, as previously stated, should be inserted to give the proper slant to the vertical side for obtaining the necessary clearance. This, of course, will throw the bevel out of true so that the wheel, if fed down close enough to it, would grind at the bottom of this surface before touching the top or cutting edge. To overcome this, the working face or upper end of the forming tool should be raised enough so that when a test indicator, attached to the wheel guard, is run along the beveled surface, the pointer will remain stationary. When finishing sides or vertical surfaces, the table should be traversed back and forth by hand, because a good surface cannot be obtained when the power feed is used, on account of the jar of the mechanism when it reverses. When grinding ordinary forming tools, little difficulty will be met with if the mechanic adheres to the use of the grain and grade of wheels recommended in the foregoing, and bears in mind that this is an operation wherein the old saying, "haste makes·waste" is especially true.

Grinding Curved Surfaces. — The making of forming tools for producing irregular curves and surfaces is quite a different proposition from that of making tools having only straight and angular surfaces and here is where experience in grinding tools is required, because these curves are affected by the clearance angle; hence, the curve on the wheel is not exactly the same as the curve to be reproduced on the work. Obtaining the required shape and size, depends largely upon one's ability in dressing the wheel to the correct curvature by hand. For instance, when grinding a tool which is to have a circular cutting edge, the wheel is first dressed to fit a true gage of the required radius and when finish-grinding, the wheel face is slightly altered by hand dressing to compensate for the angularity of the cutting face. In other words, it is necessary to make the wheel face slightly elliptical in order to secure a circular cutting edge along the beveled face at the top of the tool.

In grinding circular surfaces, it is advisable to never work on more than a quarter of a circle at a time because the work can be done more rapidly and better by following this rule. For example, to grind the tool illustrated at G in Fig. 8, the wheel should be dressed off so as to form a curve extending a little beyond the center at the bottom, as indicated by arrow a, to allow for joining the two halves of the surface neatly and smoothly. When grinding a curved surface of this kind, first grind one side and then note the dial reading on the machine. Next turn the holder around and grind the other side, feeding the wheel down to the same depth as shown by the previous reading. It is also necessary to feed the wheel laterally somewhat in order to compensate for the angularity of the top or working face of the tool. The form of the wheel must also be changed slightly in order to secure a true circle at the cutting edge.

As an example, illustrating the grinding of curves of large radii, let us assume that a tool is to be ground and that the radius of curvature required on the work is $\frac{17}{32}$ inch. Secure a partly worn 6- by 1-inch cup wheel of 46-I grain and grade and one that is a little wider than half the width of the curve. Dress off the wheel to fit a $\frac{17}{32}$-inch radius gage and then grind one side of the curve as indicated by the diagram G, Fig. 8, just cleaning up the side. Next raise the wheel and turn the fixture around for grinding the other side. Move the wheel over, and down, gradually, to the same reading as shown by the dial on the handwheel of the machine. Doubtless this will not make the rough planed recess either deep or wide enough so that the wheel must be fed down, say 0.002 or 0.003 inch, and laterally proportionately. When making these adjustments the dial readings should be noted. Now raise the wheel slightly and turn the fixture around again; then by moving the wheel over and down, the first side should be ground until the dial registers the same as for the other half of the curve. This operation must be repeated until the required depth is obtained, care being taken to keep the center of the curve in the proper place. The wheel must be dressed off at intervals for grinding the proper shape so that the cutting edge in the plane of the working face of the tool will be a true radius. All gages or templets for irregular surfaces and curves should be made to correspond with the curve required on the work and be held in the same plane as the working face of the tool, or, in this case, on a slant of 12 degrees, which is the required angle of clearance. No allowance should be made on the gage, the latter corresponding to the measurements required on the work and the tool being ground until the gage accurately fits the cutting edge, when held parallel with the top or working face.

While the wheel should be dressed to the required form with a diamond tool, the following trick is often worth knowing: After the curve on the work is ground approximately to shape and size, secure part of a broken 46-K emery wheel and using a flat surface, run this piece lightly back and forth once or twice over the working face of the wheel. Coarse emery cloth may also be used. As this operation dulls the wheel face, it must not be done until the tool is ground almost to the finished size; moreover, one must not attempt to secure the correct shape by dressing the wheel in this way. As previously mentioned, the shape should be obtained by dressing with the diamond tool, and not more than 0.001 inch be taken off of the tool after the emery wheel is smoothed with the emery brick. If there is a greater allowance, the tool is liable to be burned, when grinding with the dulled face of the wheel. When grinding curves, the wheel, as it does not feed laterally as in flat grinding, leaves fine scratches, the coarseness of which depends on the grain of the wheel; therefore a piece of emery wheel or cloth is used on the wheel prior to the final finishing operation, merely to obliterate these scratches. Ridges, however, cannot be ground out with a wheel that has been dulled in this way.

Wheels to use for Grinding Curved Surfaces. — When grinding curved surfaces, the grade of the wheel should be varied according to the radius of the curvature. For all radii above $\frac{1}{4}$ inch, the use of a 46-H wheel is to be recommended. Although this grade is a little harder to shape than a softer grade, and

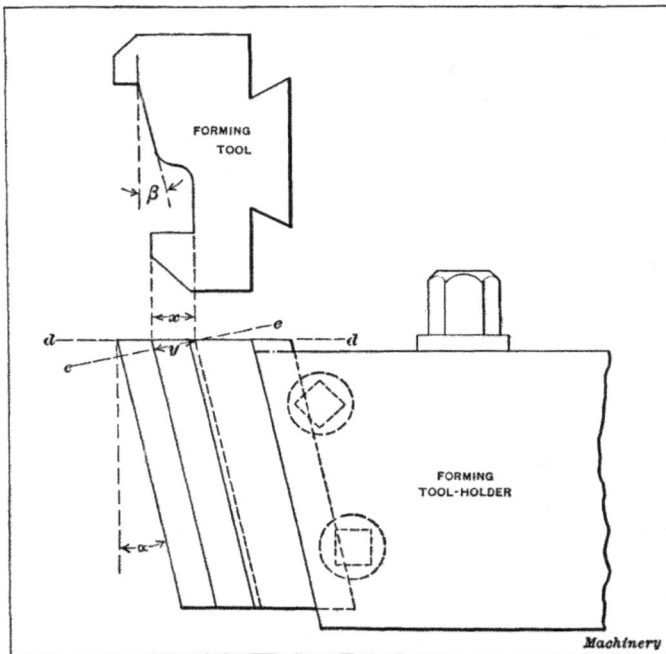

Fig. 11.   Elevation and Plan of Forming Tool showing how Clearance Angle affects Dimensions of Tool

secure a smooth surface, it will cut free without glazing or losing its form. For comparatively small radii, however, this grade will not hold its shape very well, and a 46-K is much more serviceable, down to a $\frac{1}{32}$-inch radius. For a $\frac{1}{32}$-inch or a $\frac{1}{64}$-inch radius, No. 46 grain is too coarse for good results and a grain and grade of 60-K should be employed. For rapid grinding, 46-H wheels should be used on all broad surfaces such as those illustrated at C, F, I and Q, Fig. 7. This grain and grade of wheel should also be used for grinding sides or vertical surfaces whenever a sharp corner is not required, and even then it is well to rough-grind with this wheel, finishing with a finer wheel. A grain and grade of 46-K gives a good finish when a small radius is to be left in a corner, but if a sharp corner is necessary, a 60-K wheel should be used, care being taken to keep it from glazing. If the wheel is touched frequently with the diamond tool, it will last longer than if allowed to glaze over excessively.

Vertical Forming Tool Calculations. — The vertical forming tool, as used on the Gridley automatic, is held on a 12-degree angle for clearance, but when grinding, the tool is held horizontally (as shown in Fig. 6) so that it is necessary to make an allowance on all radial as well as angular measurements, because there is a difference in the distance from one step to another on the tool, or a slight change in the angle, depending upon whether the measurement is taken along the top face of the tool or at right angles to the front face. To illustrate, if dimension x, Fig. 11, represents the distance that is to be reproduced on the work, it will be necessary, when grinding the tool, to make the depth y, measured at right angles to the front face, slightly less than the required depth x, owing to the angular position of the tool. Evidently the difference between x and y will depend upon the clearance angle α. The depth of y will equal dimension x multiplied by the cosine of angle α, or the side adjacent equals the hypotenuse times the cosine of α. For example, the part made with the tool shown at C, Fig. 7, should have a radius difference of one inch from the body part to the outside of the flange produced by the slot in the tool. As the top face of the tool is at a 12-degree angle while grinding, the slot should be ground to a depth equal to the radial dimension required on the work, times the cosine of 12 degrees, or $1 \times 0.9781 = 0.9781$ inch, instead of one inch.

Allowance must also be made when grinding angular surfaces. For example, suppose the angular surface of the forming tool illustrated in Fig. 11 is to reproduce an angle β of 15 degrees on the work. It is evident that in order to grind the tool to an

angle β of 15 degrees on the top or along the cutting edge, the
angle at right angles to the front face of the tool must be slightly
less, owing to the clearance angle. For instance, if angle β is
to be 15 degrees when measured in the plane d–d, the angle
when measured along the plane e–e (which is square with the
front face) will be a little less than 15 degrees and will represent
the angle to which the face of the grinding wheel should be
trued.

To determine the angle at which the tool must be ground,
multiply the tangent of the angle required on the work, by the
cosine of the clearance angle; the product equals the tangent
of the angle when measured at right angles to the front face of

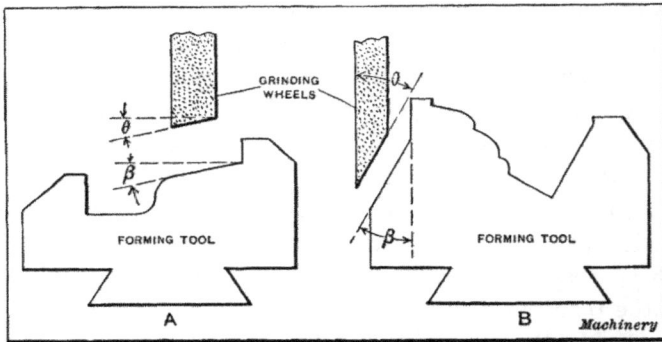

Fig. 12. The Angle of the Grinding Wheel must be varied slightly
owing to Clearance Angle of Forming Tool

the tool. Thus, taking the tangent of 15 degrees, which is
0.26795, and multiplying by the cosine of 12 degrees, or 0.97815,
we have 0.26795 × 0.97815 = 0.26209, which is the tangent of
14 degrees 40 minutes, nearly; therefore, the grinding wheel
should be dressed to this angle.

For angles larger than 45 degrees, it is preferable to work
from a vertical line, or from a line at right angles to the axis of
the work, instead of parallel to the axis. When working from
a vertical line, divide the tangent of the required angle by the
cosine of the angle of clearance, instead of multiplying, in order
to obtain the tangent of the angle measured at right angles to
the front face of the tool. The difference between the two
methods is illustrated by the diagrams A and B, Fig. 12. When

angle β is measured in the plane of the top face of the tool and from a horizontal line parallel to the axis of the work (as at A), angle θ of the wheel face equals tan β $\times$ cosine of clearance angle. On the other hand, when angle β is measured from a vertical line at right angles to the axis of the work (as at B), tan θ of the wheel face equals $\dfrac{\tan \beta}{\text{cosine of clearance angle}}$. Suppose angle β (sketch B) is 30 degrees; then, dividing the tangent of the required angle by the cosine of the clearance angle, we have: $\dfrac{0.57735}{0.97815} = 0.59024$, which is the tangent of the angle the wheel should be trued to, or approximately 30 degrees 33 minutes. Thus, we have the following very simple rules for obtaining grinding angles or the angle of the tool measured at right angles to the front face:

When working from a vertical line (at right angles to axis of work) the tangent of the grinding angle equals the tangent of the angle of production (angle required on work) divided by the cosine of the angle of clearance. When working from a horizontal line (parallel to axis of work) the tangent of the grinding angle equals the tangent of the angle of production, multiplied by the cosine of the angle of clearance.

Circular Forming Tool Calculations. — To provide sufficient periphery clearance on circular forming tools, the cutting face is off-set with relation to the center of the tool a distance C as shown at B, Fig. 1. Whenever a circular tool has two or more diameters, the difference in the radii of the steps on the tool will, therefore, not correspond exactly to the difference in the steps on the work. The form produced with the tool also changes, although the change is very slight, unless the amount of off-set C is considerable. Assume that a circular tool is required to produce a part having two diameters as shown. If the difference D_1 between the large and small radii of the tool were made equal to dimension D required on the work, D would be a certain amount over-size, depending upon the off-set C of the cutting edge. The following formulas can be used to determine the radii of circular forming tools for turning parts to different diameters:

Let R = largest radius of tool in inches;

D = difference in radii of steps on work;

C = amount cutting edge is off-set from center of tool;

r = required radius in inches;

then:

$$r = \sqrt{(\sqrt{R^2 - C^2} - D)^2 + C^2}. \qquad (1)$$

If the small radius r is given and the large radius R is required, then

$$R = \sqrt{(\sqrt{r^2 - C^2} + D)^2 + C^2}. \qquad (2)$$

To illustrate, if D is to be $\frac{1}{8}$ inch, the large radius R is $1\frac{1}{8}$ inch, and C is $\frac{5}{32}$ inch, what radius r would be required to compensate for the off-set C of the cutting edge? Inserting these values in Formula (1):

$$r = \sqrt{(\sqrt{(1\tfrac{1}{8})^2 - (\tfrac{5}{32})^2} - \tfrac{1}{8})^2 + (\tfrac{5}{32})^2} = 1.0014 \text{ inch.}$$

The value of r is thus found to be 1.0014 inch; hence the diameter = $2 \times 1.0014 = 2.0028$ inches instead of 2 inches, as would have been the case if the cutting edge had been exactly on the center-line of the tool.

Most circular forming tools are made without top rake, that is, the cutting edge is in a horizontal plane, as illustrated in Fig. 1. Tools made in this way are especially adapted for cutting brass but may not work entirely satisfactorily on tougher and harder metals. The amount of top rake for circular tools varies from 0 to approximately 18 degrees. When a circular forming tool is given top rake, corrections for the diameters of different steps must be made, the variation in diameter depending upon the angle of rake. The rather lengthy mathematical calculations involved are given in full in MACHINERY's Reference Book, No. 101, "Automatic Screw Machine Practice," page 20.

Grinding Notch in Circular Forming Tool. — The drawings A, B and C, Fig. 13, illustrate methods which are commonly used for making circular forming tools. The two saw cuts opposite the working side of the tool are intended to relieve the strains due to hardening. Before cutter A can be used the portion indicated by the dotted lines must be ground off. The method described in the following is intended to facilitate this operation. The formed

surface of a circular tool must be polished to remove the scale
if good results are desired.   For this reason most forming tools
are made as shown by Sketch A, and after the cutter is hardened
and polished, a gash, as shown at B, is ground into the cutter,
after which the cutter is ground as indicated at C.   A wide tool
made in this manner will require considerable time for grinding
in order to remove the metal without drawing the temper of the
tool.   On the other hand, if the tool was originally milled before
hardening, as shown at C, it would be almost impossible to polish
the tool after hardening.   The simplest and most effective means
for overcoming this difficulty is indicated at D; the cutter is
made with an extra slot, as shown, and after the tool is hardened
and polished, the piece y is easily broken out, without injury to

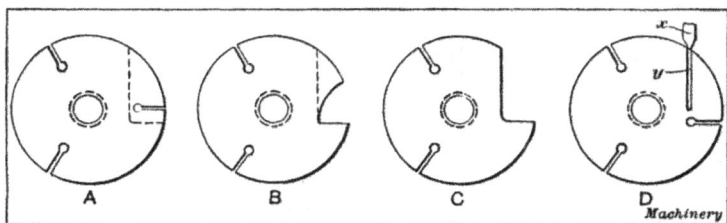

Fig. 13.   Methods of Cutting Notch in Circular Forming Tool

the cutter, by driving in a wedge as at x.   The cutter may be
held in a bench vise and only a slight tap of the hammer is necessary, after which the cut can be quickly finished off with an emery
wheel.   This method will save considerable expense in the making
of circular forming tools.

Making Concave Forming Tools in Milling Machine. — Fig. 14
illustrates a method of making a concave forming tool such as is
used for backing off convex milling cutters.   This tool has, of
course, the same shape for its entire depth so that it may be
ground and reground without changing its original form.

In the illustration, B represents the tool which is held in the
holder A at an angle of 76 degrees with the table of the milling
machine, this giving to the tool the proper cutting clearance.
The first thing to be done, after placing the tool in the holder, is
to mill off the top of the tool so that it will be parallel with the
table of the machine.   A semi-circle of the desired radius is then

drawn on the back of the tool, and with any cutter that is at hand, it is milled nearly to the mark, care being taken not to go below it. For finishing the cutter, a plug C is made, the end being hardened and ground in a surface grinder. This plug is held in a special holder D, which fits the spindle of the milling machine, and when it is set so that its axis is perpendicular to the tool, the spindle of the machine should be firmly locked. Now,

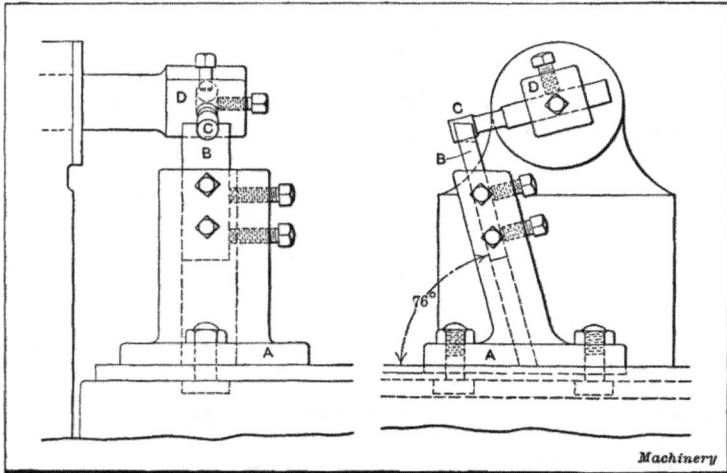

Fig. 14. Making a Concave Forming Tool in the Milling Machine

by moving the platen of the machine backward and forward by hand, the plug can be made to cut a perfect semi-circle in the tool.

It is good practice to plane a little below half of the diameter of the plug, thus allowing some stock to be ground off after the tool is hardened. In hardening, these tools usually come out very satisfactorily, but, if any distortion takes place, it will be from the sides, and may be readily remedied by a little stoning. By using the concave tool for a planing tool, a convex tool may be formed, but in doing so care should be taken that both tools stand at an angle of 76 degrees with the bed of the machine. This shape of tool would be used for backing off a concave cutter.

Making Forming Tools for Gear Cutters. — The making of a forming tool for forming gear cutters is not work that is frequently given to the toolmaker, but there are some instances when a special gear cutter is wanted and lack of time prevents having it made by the cutter manufacturer, who is not always very prompt in filling such orders. The ordinary practice is to lay out the tooth curve full size or several times enlarged, and then make master tools to this drawing, by which the formed tool is planed. Following this method, it is necessary to make a special master planing tool for each side of the tooth and planing tools for the

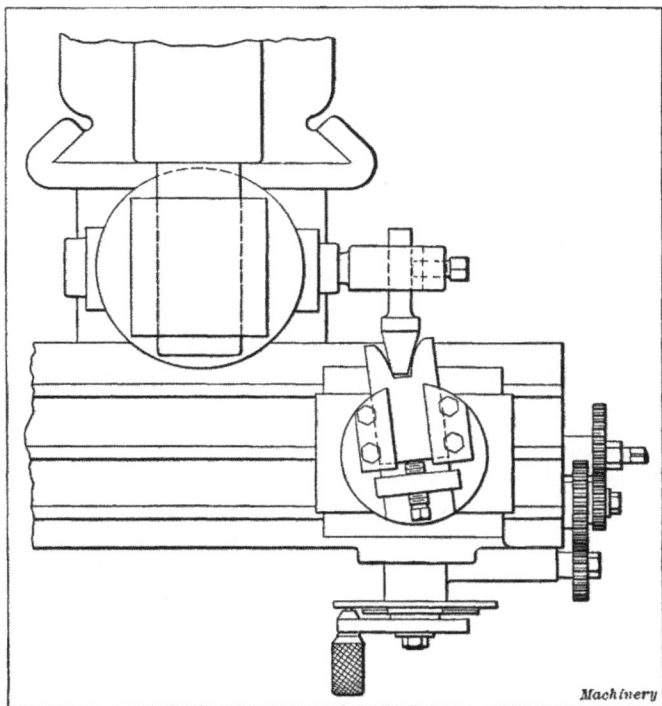

Fig. 15. Milling Machine set up to generate Forming Tool for Gear Cutters, with a Rack-shaped Tool

curves at the bottom; these operations combined with the method of using the tools and making the necessary corrections for the distortion due to the angle at which each tool is set in planing, are all likely to lead to inaccuracies in the tooth curve, making the results anything but satisfactory. By the following method (which was described by John Edgar in MACHINERY, July, 1914), all the multiplied inaccuracies incident to copying from this original draft are omitted. This method consists of generating the forming tool direct without any intermediate steps, and by mechanical means that alone determine the shape of the tooth curve, so that a correct involute curve is obtained without any approximations whatever.

Probably the greatest drawback to the first method of procedure referred to, is the necessity of making the original drawing of the tooth curve, there being so many methods in use that are mere approximations. Such methods are all well enough for the purpose of representing gear teeth on a drawing, but if it is attempted to make the teeth themselves to these layouts, trouble is liable to be the result. Almost any toolmaker worthy of the name can produce a tool very close to the original curve; but the proposition of getting exactly the original curve is a job that must be left to some mechanical means if we are to get the closest possible degree of accuracy. This means is found in every toolroom and is the universal milling machine. For the operation of generating the tool, a universal milling attachment is required, which may make the use of the method impossible in some cases; but these attachments are now commonly found in the toolroom in connection with the milling machine.

Generating Forming Tool With Fly Cutter. — Fig. 15 shows the machine arranged for using the rack-shaped fly cutter. The forming tool is clamped in a special chuck held in the spindle of the dividing head, which is set in the vertical position. The head is shown set up with the change gears as for spiral cutting, the pitch for which the gears are chosen being equal to the circumference of the pitch circle. The universal spindle of the attachment is set at right angles to the axis of the main spindle and parallel to the direction in which the table travels. When set up in this manner, the machine is ready to generate the involute sides of the forming tool, by using a fly cutter that is shaped like the tooth of a rack of the same pitch. The process is similar to that used in some way in all generating machines. By setting the top of the forming tool blank at the height of the axis of the cutter spindle, and the fly cutter so that the cutting plane passes through the center of the cutter spindle, next adjusting the fly cutter to produce a gash of the proper depth, and then starting the machine and throwing in the feed for the table, we can generate the tooth space shown. This space is of the correct shape at the top of the tool only, as the angle of the fly cutter with the top of the forming tool affects the shape at all other positions deeper in the tool. A tool made in this manner would appear as shown at *A* in Fig. 16, and to make it of use as a forming tool it must be relieved up to the top for clearance. This can be done by filing or other means, so that the finished tool would have a 20-degree clearance (as indicated by the dotted line at *B*), extending all around the form. With the fly cutter made to close dimensions, the space generated will be of the correct width at the pitch line and to depth; and a forming tool made with it would likewise be of the correct size in relation to thickness and depth.

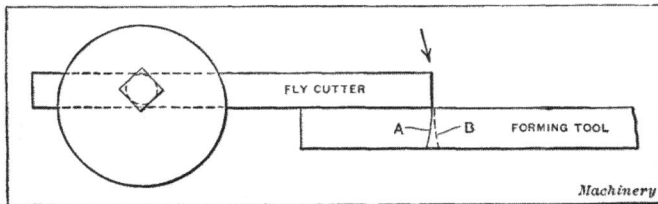

Fig. 16.   Shape of Tool produced by Fly Cutter when Top of Blank is at Height of Cutter Spindle Axis

The necessity for relieving the tool after generating may be avoided by setting the fly cutter as shown in Fig. 17, 20 degrees being chosen as giving plenty of clearance for the tool in the forming and backing off of the cutter. If the cutter teeth are to be given more than the ordinary amount of relief, the angle should be increased. With the use of the fly cutter in this position, a correction will have to be made to counteract the angle at which it is set. This is done by making the angle of the fly cutter such that the angle on a line parallel with the top of the forming tool will be twice the pressure angle. The corrected angle of the side of the fly cutter may be found by the formula:

$$\text{Tangent of angle of side of fly cutter} = \frac{\sin \alpha}{\cos \beta}$$

where α = pressure angle and β = clearance angle.

The included angle of the tool is twice the angle of the side, and the height at which the forming tool is set above the center of the cutter spindle is found by trial, by setting the fly cutter as shown in Fig. 17 and bringing the forming tool to the height of the tip of the fly cutter. This height is, of course, dependent on the radius of the circle swept by the fly cutter. The fly cutter should be set to sweep as large a circle as possible, to give the least possible amount of concavity to the forming tool. The shape generated by the fly cutter is nearly correct throughout the thickness of the forming tool when the latter is thin as compared to the sweep of the fly cutter; and the forming tool can be sharp-

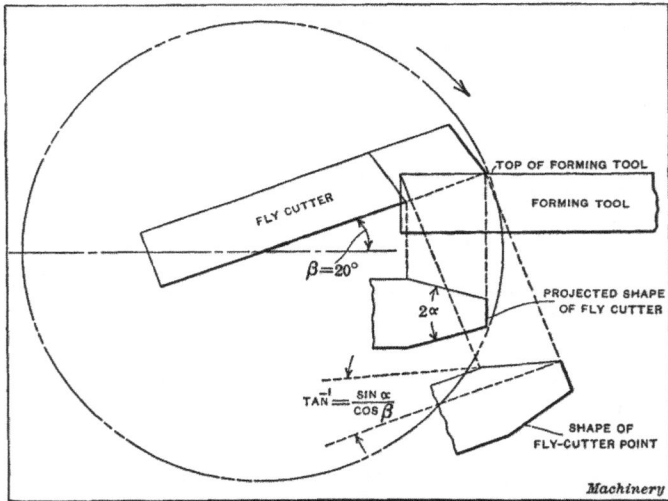

Fig. 17. Method of Setting Fly Cutter to Obtain the Required
Clearance on the Forming Tool

ened by grinding across the top face without greatly changing its form. The clearance, as produced in the case of the fly cutter set according to the preceding instructions, is not adapted to work that requires side relief, as in the case of small numbers of teeth, nor for bevel gear cutters.

This method of generating a forming tool is especially valuable when it is desirable to make cutters for any special number of teeth, when a standard cutter is not satisfactory.

CHAPTER IV

PRECISION THREADING

The principal requirements for cutting an accurate thread in the lathe are an accurately made thread tool, a correct setting of the tool, and a lathe with an accurate lead-screw. In making a U. S. standard thread tool a correct 60-degree angle gage is necessary. To produce such a gage, first plane up a piece of steel in the shape of an equilateral triangle as shown at *a* in Fig. 1. After hardening this triangle, grind and lap the edges until the three corner angles prove to be exactly alike when measured with a protractor. This is the master gage. To produce the female gage, make two pieces, one right- and one left-hand, like that shown at *b*; harden them and lap the edges that form the 150-degree angle so that they are straight, and square with the sides. When this is done the two pieces should be screwed and doweled to a backing plate *d* as shown, using the master triangle to locate them, thus producing an accurate female gage.

In making up the tool, some form of cutter to be used in a holder should be chosen in preference to a forged tool on account of convenience in handling and measuring and the ease with which it may be re-ground without destroying the shape. The tool should be made so that the top will be level when in the holder, and the clearance should be about 15 degrees, which is ample for a single thread unless the pitch is very coarse. With that amount of clearance, the included angle between the sides of the tool in a plane perpendicular to the front edge is approximately 61 degrees 44 minutes. The tool should be planed to that angle as nearly as is possible by measuring with a protractor; then, to test its accuracy, it should be placed top down on a flat piece of glass *c* and tried with the 60-degree gage as shown in Fig. 1. After lapping the tool until it shuts out the light when tried in this manner, the angle may be considered as nearly correct as it is possible to obtain with ordinary means.

To adapt the V-thread tool thus made to cut the U. S. standard form of thread, it is only necessary to grind off the sharp point an amount equal to one-eighth of the depth of a V-thread of the required pitch. The depth of a V-thread equals the constant 0.866 divided by the number of threads per inch. Therefore, if there were, say, 20 threads per inch, the amount to grind off would equal $\dfrac{0.866}{20} \times \dfrac{1}{8} = 0.0054$ inch. To test the accuracy of this grinding, a piece of steel should be turned up to the correct

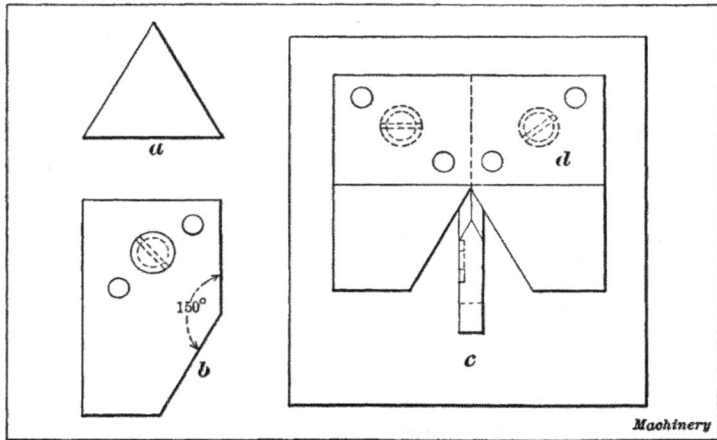

Fig. 1.   Gages for Testing Threading Tools

outside diameter of the thread and a short shoulder turned down at the end to the correct diameter at the bottom of the thread; then the piece is threaded, the tool being fed in until the flat of the tool just grazes the shoulder.   Then cut a nick in the edge of a piece of sheet steel with the threading tool. This sheet steel piece is now applied like a gage to the threaded cylindrical piece. If the nick in the sheet steel fits the thread so that it shuts out the light the flat of the tool is correct.

Making Accurate Tools for Whitworth Thread. — While the development of a correct tool for a U. S. standard thread or a V-thread requires skill and patience, it is easy compared to the task of producing a tool for a thread having a round top and bottom like the Whitworth and British Association standards. When testing the accuracy of threads of this type, they are not only measured by gages and micrometers, but the curves must match the angle so evenly that when the male gage is tried in the female from either end, no difference can be detected. It may be laid down as a cardinal principle that the best results are obtained by developing the form first with a flat top and bottom, as in the U. S. standard thread, rounding the corners afterward. The first step of all is to produce a correct angle gage; assuming that we are to make a Whitworth thread tool, a gage measuring 55

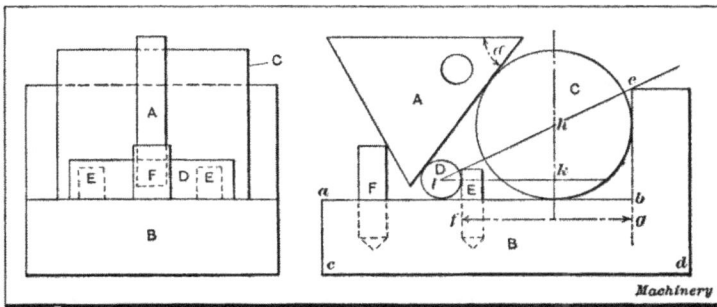

Fig. 2.   Method Employed for Obtaining 55-degree Angular Gage

degrees is required.   Make and harden a steel triangle A, Fig. 2, with the angle α made as near 55 degrees as is possible by using a bevel protractor; the other two angles are to be equal.   Then make an angle iron B with sides ab and cd parallel, and be square with ab.   Assuming that C and D are accurate 2-inch and $\frac{1}{2}$-inch plugs, insert the pins EE in such a position that a line drawn through the centers of C and D, at right angles to their axes, will make an angle of $27\frac{1}{2}$ degrees with ab.   This can be done by figuring the distance fg as follows: in the triangle lhk, $hk = 1 - 0.25 = 0.75$ inch.   Then the side $lk = 0.75$ cotan $27\frac{1}{2}$ degrees = 1.4406 inch.

$1.4406 + \frac{1}{2}$ diameter of $C - \frac{1}{2}$ diameter of D =
$1.4406 + 1 - 0.25 = 2.1906$ inches $= fg$.

Set the pin F near enough to D to keep the corner of the triangle from striking the angle iron B.   Mount the triangle A as shown,

and set up the fixture on a surface grinder table using a toe strap
or finger clamp in the small hole in *A* to hold it in position, and
grind first one edge, and then the other.   In this way the male
angle gage is produced.   A female gage can be made to this male
gage by the method described previously for U. S. standard
thread gages.

The tools to be used in making Whitworth thread tools are
shown in the upper part of Fig. 3.   They include an angular tool
with a flat point, the width of the point being such that it reaches

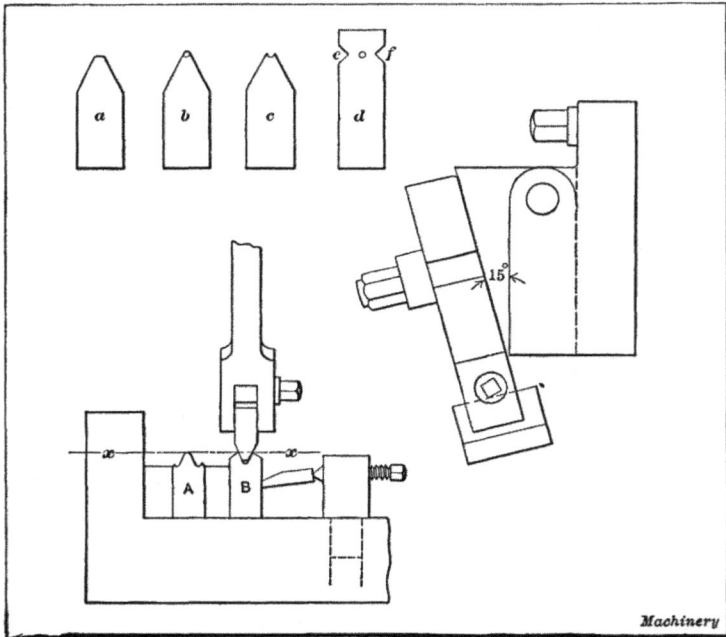

Fig. 3.   Method of Planing Whitworth Thread Tools — Tools used

to the center of the round in the bottom of the thread, and the
angle matching the gage previously made; a female radius tool
for forming the point; and a male radius tool for the side radii.
For convenience in measuring and getting the exact form required, these tools should be made with the top square with the
face at the cutting edge, *i.e.*, without clearance.   The sides and
back of all should be ground as well as the top.   The tool *a* can

be ground by means of an angular block made in the same manner as the male angle gage and should be finished by lapping. The tool *b* can be made in two pieces, one a hardened, ground, and lapped wire, and the other a soft piece made up in such shape that the wire can be soldered or otherwise firmly fastened to it in the correct position. The tool *c* should be made up first as at *d* and hardened. Then lap the hole carefully to size and grind the outside. After measuring the distance from the hole to the back of the tool, the front can be ground off to *ef* and the bevels ground until the depth of the round part is right.

We now require a shaper with an apron arranged to hold the tool-holder at an angle of 15 degrees, as shown in Fig. 3. The apron should fit the clapper-box perfectly. If it does not, it is better to fasten it solid, and let the tools drag back through the cut, sharpening the tools over again before finishing; otherwise, one runs the risk of side shake. With this angular apron we can use the tools made without clearance to produce a tool with correct clearance for the lathe. Two thread tool blanks, one, *A*, of tool steel, and one, *B*, of machine steel, should be set up on the table adapter as shown in the illustration, with spacing parallels between to avoid interfering with one while planing the other. The blanks should be planed off to exactly the same height, and all measurements for height should be figured from the line x–x, allowance being made for the difference caused by the 15-degree clearance. Then, after carefully measuring the tools previously made, to determine where the exact center is, we can start forming the blanks, setting the tools sidewise successively by positive measurement from the side of the adapter. The angular tool is used first and with it we plane down the sides of the tool *A* and the center of *B* so that the point of the tool just reaches the center of the radius. Then using the female radius tool, round the point of *A* and the two points of *B*, feeding down until the circle of the tool is just tangent to the top of the blanks. The male tool will round out the two lower corners of *A* and the center of *B*, being fed down to the exact depth.

We now have the thread tool *A*, which can be hardened, and the machine steel blank *B* is used as a lap to correct errors in tool

A, reversing the lap occasionally, and using oilstone powder or other fine abrasive as the cutting medium.   Great care must be used in putting on the abrasive, as in all lapping operations of this kind points and corners are liable to lap faster than wide surfaces.   This operation does not really correct the tool, but equalizes the errors due to imperfect matching of the different cuts, and it can be done so effectively that whatever errors of that kind are left cannot be detected.

To test the tool, turn up a blank plug with a teat equal to the diameter at the bottom of the thread.   When this is threaded, the point of the tool should just touch the teat, when the outer corners touch the top of the thread.   The angle of the thread may be given a practical test by measuring the diameter of the test plug by the use of wires and a micrometer.   (See "Three-wire System of Measuring Threads.")

For the final test of the fit of the curves with the angle, a tap should be threaded with the tool, and a female gage tapped with the tap.   The plug just made must screw into this with an equal amount of friction from either end, and show a full contact on the thread.   If this last test is not successful it shows that the lapping is not good enough and must be done over.   If the plug does not measure right, it is necessary to go back to the planing and plane up another tool, making such allowances as one judges will correct the error.   It is sometimes necessary to do this several times before a perfect tool is produced.   In the use of the tool in the lathe, great care is necessary to see that it is set parallel with the center of the spindle, and so that the two side curves will scrape the top of the thread at the same time. With the exception of making the angle gage and tool grinding block, this whole procedure has to be carried out for every pitch required.

Measuring Flat on Acme and U. S. Standard Thread Tools. — Occasionally the commercial gages for testing Acme and U. S. standard thread tools do not have notches of the required pitch and cannot be used.   The diagrams shown in Fig. 4 illustrate two methods of measuring the width of the point or "flat" which give a close approximation of the actual value.   The method shown to the left is based on the fact that the spindle of a micrometer is of known diameter. A scale or parallel is laid on the spindle and the micrometer anvil, the tool being placed with its flat resting against the scale. The micrometer is then closed to the position shown, and 0.2887 inch is subtracted from the reading in the case of a U. S. tool, and 0.1293 inch for an Acme or worm thread tool.

The values 0.2887 and 0.1293 are used only when the micrometer spindle has the usual diameter of 0.25 inch. To obtain the value or constant that must be subtracted from the reading, multiply twice the spindle diameter by the tangent of one-half

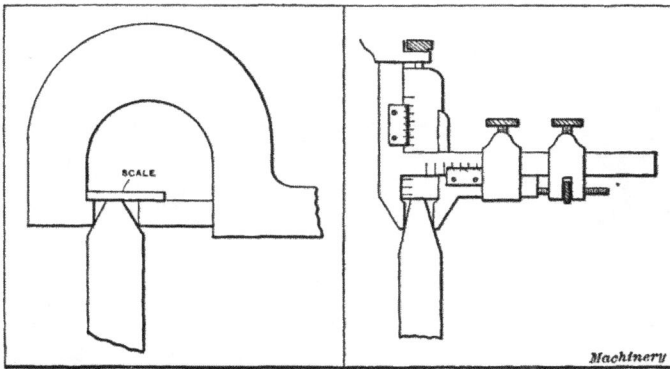

Fig. 4.   Two Methods of Measuring the Flat on U. S. Standard
and Acme Thread Tools

the thread tool angle. Thus for a U. S. standard tool the amount equals 2 × 0.25 × tan 30 degrees or 2 × 0.25 × 0.57735 = 0.2887, approximately.

The sketch to the right in Fig. 4 shows the method of measuring the flat with a gear tooth caliper. If the measurement is made at a vertical distance of $\frac{1}{4}$ inch from the point, the same values, i.e., 0.2887 inch for U. S. standard and 0.1293 inch for an Acme thread tool, are subtracted from the readings of the caliper to obtain the actual width of the flat. For example, if the caliper reading was 0.315 inch, the width at the point of an Acme thread tool would equal 0.315 − 0.1293 = 0.1857 inch, assuming that the sides were ground to the standard angle of 29 degrees. The width of the tool point, or the width of the

flat at the bottom of an Acme thread, should equal 0.3707 ×
pitch − 0.0052 inch.   The width of the tool point or flat for a
U. S. standard thread equals one-eighth the pitch.

At *A*, in Fig. 5, is shown a special gage for measuring the flat
on U. S. tools, and at *B* a gage for making the same measurement on Acme tools.   In both cases, the reading of the micrometer from outside to outside of the pins *b* and *c* is exactly one
inch greater than the width of the flat.   In using these gages,
the thread tool is held with its point resting against the straightedge *a* and its sides between the fixed jaw *d* and the sliding
jaw *e*.   The distance from outside to outside of pins *b* and *c* is

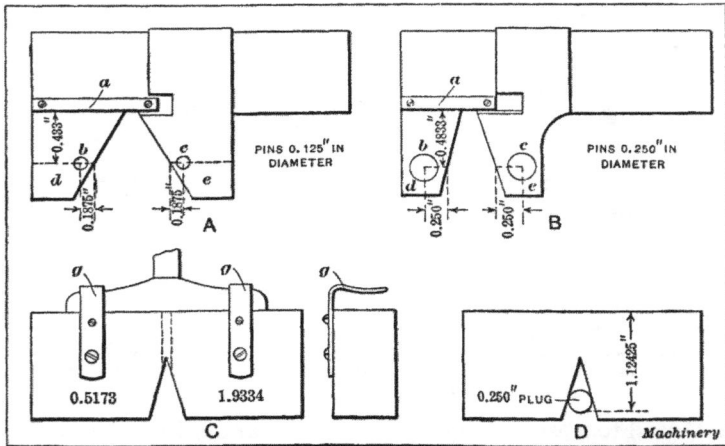

Fig. 5.   Special Gages for Testing Width of Flat on U. S. Standard
and Acme Thread Tools

then measured with a micrometer and this measurement is one
inch greater than the width of the flat.   If these gages are
carefully made according to the dimensions given, their use will
enable the measurement of U. S. and Acme thread tools to be
accurately made.

A tool capable of making these measurements even more
accurately is illustrated at *C*.   The gage consists of a block of
steel of about $\frac{5}{8}$ by $1\frac{1}{8}$ by $3\frac{1}{4}$ inches in size, having a 29-degree
V-groove in one face.   A hole is drilled through the block from
the opposite face to the vee, and two springs *g* are provided to

hold the micrometer depth gage in the position shown. In making this gage, the upper face is machined down until the distance from a 0.25-inch diameter plug held in the vee, to the opposite side of the block, is 1.12425 inch, as shown at D. This locates the vertex of the 29-degree angle at a distance of 0.500 inch below the face of the block. In using the tool, the depth gage rod is run out 0.500 inch, so that its end coincides with the vertex of the 29-degree angle, which represents the "zero" from which measurements are made. With an Acme tool held in the vee, the distance from this zero, as measured by the depth gage, is 1.9334 times the width of the flat. The width of the flat is obtained by multiplying the gage reading by 0.5173. It will be found convenient to have the preceding constants stamped on the front of the block and by preparing a table giving the gage reading for tools ranging from 1 to 12 pitch on the back of the block, the work of making calculations will be greatly facilitated.

Pitch.	Gage Reading.	Pitch.	Gage Reading.
1	0.7066	5	0.1334
1¼	0.5274	5½	0.1202
1½	0.4677	6	0.1094
2	0.3482	7	0.0924
2½	0.2767	8	0.0795
3	0.2289	9	0.0696
3½	0.1949	10	0.0617
4	0.1692	11	0.0551
4½	0.1493	12	0.0497

In the preceding table, a clearance of 0.010 inch is allowed. In making the tools shown at A and B, Fig. 5, the pins b and c are set so that the distance from outside to outside is one inch when the gage is closed. In the gage for U. S. thread tools shown at A, the distance 0.433 inch from the center of the pins to the top of the gage, is determined by the ratio of the base to the perpendicular of a 30-degree triangle. In this case the base of the triangle is 0.250 inch, and this fixes the perpendicular as shown in the illustration. Similarly for the Acme tool shown at B, the base of the 14 degree 30 minute triangle is 0.125 inch, and this requires the perpendicular to be 0.4833 inch. The constant 0.5173 for the tool shown at C, is obtained from the relation of the perpendicular to the base of a 14 degree 30 minute triangle. Thus the width of the flat is twice the gage reading × tan 14 degrees 30 minutes.

Making Precision Thread Gages. — It appears to be the general idea that screw plug gages must be made of tool steel, but it has been found very practical to make them of cold rolled stock, which is very soft and easy to cut, but which, when hardened, gives a surface which is fully as hard as tool steel. This hard surface extends deep enough into the thread gage to permit grinding 0.005 inch deep, enough hard surface still remaining to prevent rapid wear when in use. Another reason for using this soft steel is that it is not likely to change its shape after having been finished, as does sometimes even the best tool steel, if it has not been properly seasoned after hardening.

Method of Setting Thread Tool. — For setting a thread tool for cutting a correct thread, a cylindrical thread gage is made, as shown in Fig. 6. This thread gage has an advantage over the ordinary thread gage on the market, in that it can be placed between the centers of the lathe; consequently one does not depend upon any secondary surface against which to set the thread gage. This is the case with the ordinary thread gages, which have to be lined up either against the side of the face-plate of the lathe, or against the side of the work, and in this way small errors are almost always introduced. The thread gage in Fig. 6 is made of machine steel, hardened and ground all over. The main body, A, is provided with three grooves, having inclusive angles of 29, 55, and 60 degrees, respectively, to correspond with the Acme, Whitworth and United States standard threads. When the gage is hardened, the two sides of the grooves are ground with the same setting of the slide-rest, the piece A being reversed on the lathe centers while grinding. This insures that both sides of the angle in the gage make the same angle with the axis of the gage.

In one end of the body A a hole is drilled; this is ground until the bottom of the hole comes exactly in line with the axis or center-line of the body *A*. A hardened and lapped plug *B* is inserted into this hole and held with a set-screw, having a brass shoe at the end. The purpose of this plug *B* is to afford a means for setting the thread tool in the lathe at the correct height, or exactly in line with the axis of the spindle. This is done by merely loosening the clamp which holds the thread tool in its tool-holder, and adjusting the tool so that the upper face bears evenly on the lower side of the plug *B*, as shown in the end view; the clamp of the thread tool-holder is again tightened, and the tool is in the correct position as to height. This is a good way of setting the thread tool to the same height as the axis of the lathe centers although it does not necessarily insure

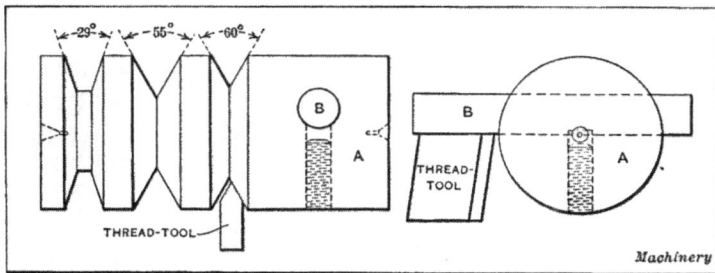

Fig. 6. Gage for Setting Thread Tools

that the thread tool, in all cases, will be set absolutely correct. If the thread tool-holder should be tipped somewhat out of the horizontal position, the top of the thread tool itself would not be horizontal, and, consequently, when the gage pin *B* was brought down upon the top of the tool, this pin would not be fully horizontal, and the thread tool would not be set to the exact height of the lathe centers. A gage or indicator should, therefore, first be used for ascertaining whether or not the top of the tool is parallel with the ways of the cross-slide. If the top of the tool is parallel, the setting obtained by the method outlined will be correct.

With the gage remaining between the lathe centers, the angular end of the thread tool is set to a correct position sideways. This setting is also a check on the accuracy of the angle of the thread tool. A piece of white paper should be placed under the gage and the tool, and a magnifying glass should be employed. The setting and the angle may be considered satisfactory when the tool fits the gage so that all light is shut off. The thread tool being set, we are now ready to proceed to finish thread the screw plug gage, which has previously been roughed out by a chaser having three or four teeth, leaving about 0.005 inch for the finishing single-point thread tool. The finishing of the thread is continued until 0.0015 inch is left for lapping. The chaser, as well as the single-point tool, should have a clearance of 15 degrees on the front face of the thread tool. This

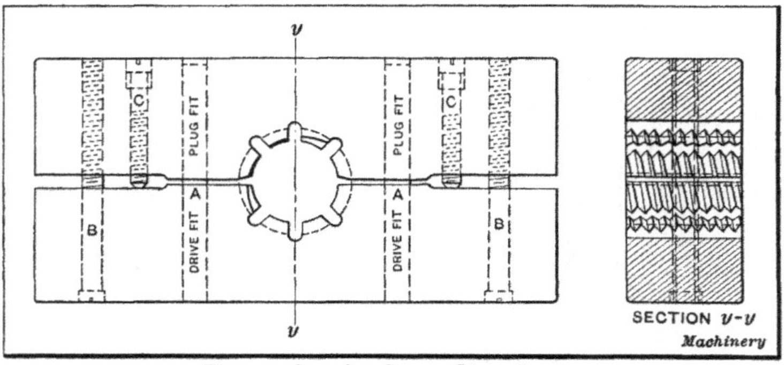

Fig. 7. Lap for Screw Plug Gages

angle has proved to be the most advantageous under ordinary conditions.

A Thread Gage Lap. — After having been finish threaded, the screw plug is casehardened and ready for lapping. A lap made as shown in Fig. 7 is used. It will be seen that this lap is somewhat different from those ordinarily used for this work. The construction shown has been adopted because of the difficulty met with in circular laps which are split on one side for adjustment, but have nothing on the sides to hold the two sections in perfect alignment; consequently, each of the sides has a tendency to follow the lead of the screw plug when lapping, and difficulty is experienced in getting a thread with perfect lead. The lap here shown, therefore, has a dowel pin *A* on each side for the purpose of holding the two sections in perfect alignment, and the adjusting screws *C* are inserted outside of the dowel pins. The two screws *B*, finally, clamp the two halves together. When the lap is assembled and screwed together, it is roughed out in the lathe with a threading tool, or tapped with three or four different sized taps, following one another in proper order. The lap is then taken apart, and planed on the inside to permit of adjustment; three grooves are cut in the thread on each side of the lap, for holding reserves of emery and oil. This will permit constant lubrication of the lap, and constant charging when lapping the screw plug to size.

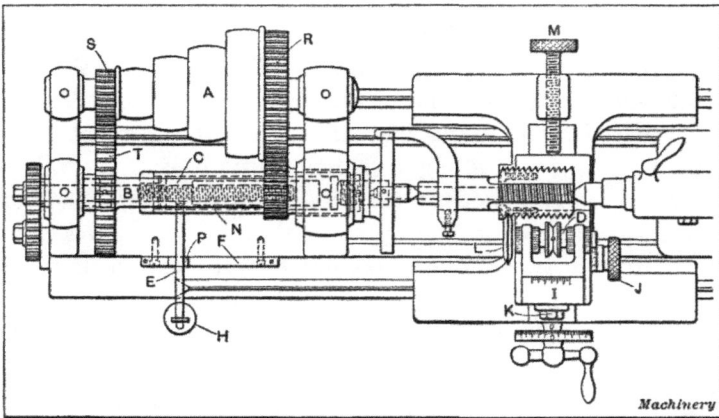

Fig. 8. Lathe Equipped with Special Mechanism for Grinding Thread of Precision Tap

Grinding the Thread of a Precision Tap. — The lap shown in Fig. 7, which is used for lapping a thread gage, as previously explained, is finished with a master tap, which must be made with extreme accuracy. This tap is ground in the angle of the thread, as shown in Fig. 8, and it is finished to a dimension 0.002 inch below the required diameter of the thread plug to be made, in order to permit the lap to wear down to the size when lapping.

The lathe must be revolved very slowly when grinding the master tap, the revolutions of the spindle being from 20 to 100 per minute, according to the size of the tap. As the illustration shows, the cone pulley is placed where the back-gears ordi-

narily are located. Gear R is disconnected, and the drive is
through gears S and T. The reason for having the cone pulley
in the back, is because the space directly under the usual location of the cone pulley in the center of the lathe is used for a
mechanism intended to permit a slight adjustment of the lead
of the tap when grinding in the angle of the thread.

The feed-screw B is placed in the center of the lathe bed,
directly under the driving spindle, and fits into a solid nut C,
from which, through the medium of a casing N and a connecting-
rod, the carriage is moved. A rod E is screwed into the nut C,
this rod extending over the side of the lathe, and resting upon
the edge of plate F, which can be so adjusted that it inclines from
o to 20 degrees. Between this plate and the rod E, a shoe P is
placed. On the extreme end of the rod hangs a weight H which
holds the rod against the plate F. This arrangement serves
the purpose of giving a slight change in the lead of the tap being ground, as it is evident that when the rod E travels along
the inclined plate F, it slightly turns the nut and moves it forward a trifle in excess of the regular forward motion imparted
to the nut by the lead-screw. By inclining the plate F in the
opposite direction, the motion of the nut may be correspond-
ingly retarded.

A grinding fixture I fits the slides on the top of the carriage.
On the right-hand side of this fixture is placed a knurled handle
J, graduated to thousandths of an inch. This handle is for the
fine adjustment of the fixture, enabling the grinding wheel to
be set correctly to the center of the thread, before starting the
grinding operation. The top of the fixture swivels in a vertical
plane, so that the wheel L, which is made of tool steel and
charged with diamond dust, can be set at an angle to the vertical,
either to the right or the left, according to the pitch and direction of the thread. This adjustment is made by loosening the
nut K which binds the head in position when set to the correct
angle. The wheel L is provided with a shank which fits a
tapered hole in the spindle D, which latter runs at a speed of
20,000 revolutions per minute. A solid backstop M is provided
to hold the fixture securely in place while working. The lathe

spindle, with the tap, and the grinding spindle run in the same direction, the same as in an ordinary grinder.

A good supply of sperm oil should be used when grinding the tap, and it is necessary to have a cover over the wheel, to prevent the throwing out of oil. This cover, however, is not shown in the illustration. Care should be taken not to force the wheel into the work, because if that is done, the accuracy will soon be impaired. The wheel should just barely touch the work, and should be fed in a very small amount, say, 0.00025 inch at a time. A sound magnifier or "listener" should be used, to hear whether the wheel is cutting moderately.

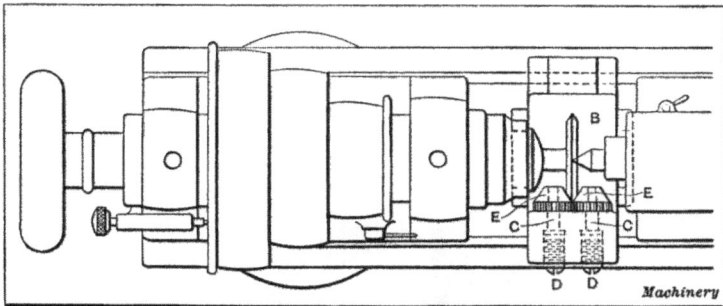

Fig. 9. Bench Lathe with Fixture for Charging Diamond Lap

The wheel is charged in the following manner: A chuck, with a tapered hole which fits the shank of the diamond wheel, is placed in the spindle of the bench lathe, as shown in Fig. 9, and the tailstock center is pushed up at the other end to provide a good support when charging. Fixture *B* is placed in the bench lathe, and clamped with a bolt and nut from underneath the lathe, about the same as an ordinary slide-rest. The front end of the fixture extends up vertically above the center of the spindle. In this projecting part, two holes are drilled, reamed, and counterbored, at the same height as the center of the lathe spindle. In these holes are fitted two studs *C* having a T-head inside the counterbored hole. Between the T-heads of these studs and the screws *D* lie fiber washers, which act as friction stops. On the other end of plugs *C* are placed hardened and ground rollers *E* having one end beveled to a 30-degree angle, while the other end has spur gear teeth milled, which mesh into each other. With the slowest speed of the bench lathe, the fixture is fed in by hand, and having two slides at right angles to each other, the same as an ordinary slide-rest, it can be located in the proper position without much trouble. A piece of soft steel wire should be flattened out to make a spade, with which to take up the diamond dust for charging the wheel. One should not try to use a piece of wood, or a brush, because they will waste the diamond dust. The master tap, which is to be ground, is relieved up to within $\frac{1}{16}$ inch from its cutting edge with a file, this being done in order to prevent any more

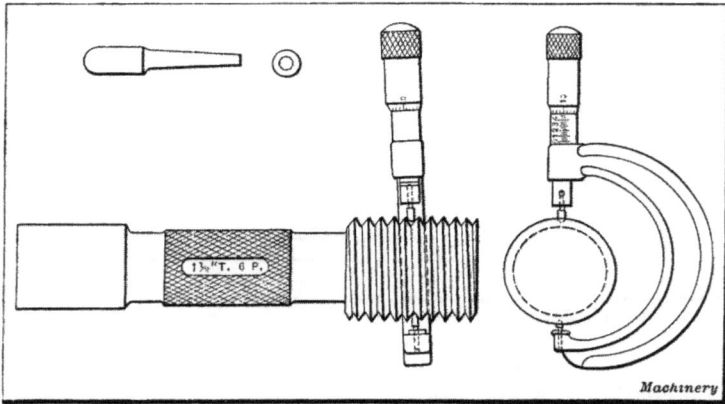

Fig. 10.   Comparing Angle Diameters with Ball-point Micrometer

grinding than is absolutely necessary, and to permit the tap to cut freely. The length of the threaded part of the master tap should be about two times its diameter.

Use of Thread Gage Lap. — The master tap being finished, the lap (Fig. 7) for the screw plug gage is tapped, and ready for use. When charging this thread lap, great care should be taken not to force the lap too much. The spindle of the lathe, where the lapping is done, should be run very slowly, with the back gears in, until the lap is thoroughly charged with emery mixed with sperm oil. Then the lathe may be speeded up to a higher speed, according to the size of the screw plug. It is poor practice to use too much emery on the lap. Reverse the lap often, and use it the same amount on both sides.   If a large number of screw plugs are to be lapped, all of the same size, lap them all, one at a time, with the lap at the same setting.   In this way the lap keeps its shape better, and can be used a long while before being retapped.   Do not attempt to tap the lap with the master tap when charged with emery, but use a roughing tap first, and also wash out the lap in benzine before tapping. When the screw plug has been lapped to within 0.0005 inch of the required size, it is ground on its outer diameter, if it be a U. S. standard thread plug, and then finished by lapping after

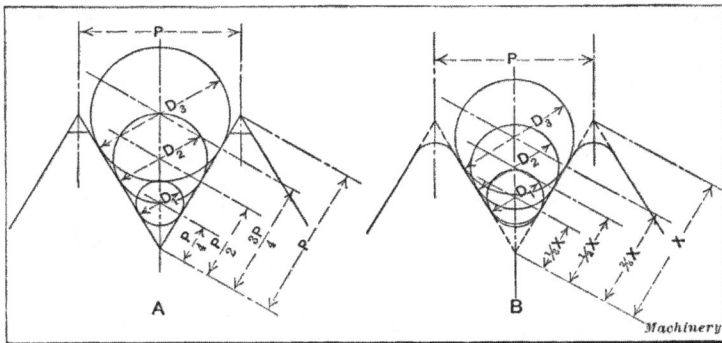

Fig. 11.   Method of Applying Ball Points when Testing Angle of
U. S. Standard and Whitworth Threads

being ground.   This will permit the top corners to be kept sharp, and better results will be obtained all around.

Gaging the Thread Angle. — Great care must be exercised during the lapping operation to see that the angle of the thread is correct.   The gaging of the angle of the thread is accomplished in the following manner: Three micrometers are used to measure the angle.   Two ball points of the same size are placed in tapered holes in each micrometer, as shown in Fig. 10.   These ball points are ground all over, and made to the shape shown in the upper left-hand corner of the illustration.   The body of these ball points is ground parallel, and then the end is turned and ground to a ball shape as shown.   Three sets of ball points are used for each pitch, one to measure the thread near its bottom, one at the center, and one near the top, as indicated at *A* and *B* in Fig. 11.   The master screw plug is used for comparison;

one micrometer is set to the master screw plug at the bottom of the thread, in the manner indicated in Fig. 10, and is then tried on the thread plug being made. The difference in diameter between the measured diameter on the master gage, and that on the plug being made, is noted. Then the two other micrometers, measuring at the center and near the top of the thread, are used, and the difference between the master gage and the screw plug diameters at the places where these micrometers measure, is also noted. If all three micrometers show the same amount of difference in relation to the master plug, then the angle of the thread evidently must be correct. After that, the micrometer

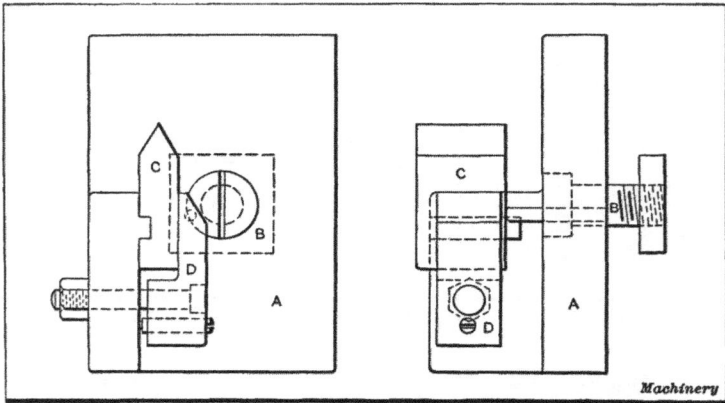

Fig. 12.   Gage for Testing the Angle of the Thread

measuring at the center of the thread is used to measure the size of the screw plug, comparing it with that of the master gage, until the plug is finished to size.

The diameters D_1, D_2, D_3, respectively, are the diameters of the cylindrical portions of the ball points used, and are, of course, also the diameters of the half-spheres on the end of the ball points. These diameters for a U. S. standard thread (Sketch A, Fig. 11) may be obtained by the following formulas: $D_3 = 0.866 P$; $D_2 = 0.5774 P$; $D_1 = 0.2887 P$. For a Whitworth thread (Sketch B), $D_3 = 0.7516 P$; $D_2 = 0.5637 P$; $D_1 = 0.3758 P$; $X = P \div 2 \sin 27\frac{1}{2}$ degrees. In these formulas, $P = $ pitch of thread.

For testing the angle of the screw plug, when finally finished to a limit of 0.0005 inch, an ordinary lathe (Fig. 13) fitted with a fixture *A*, shown separately in Fig. 12, is used. The toolpost is taken off the lathe, and replaced with this fixture, which is clamped in the T-slot of the toolpost slide, with bolt *B*. The thread gage *C* is ground all over, and the angle fitted to a master gage. The gage *C* is held by the tongue and groove on the left-hand side of the fixture, and clamped with a strap *D*. To set this gage correctly, in relation to the axis of the spindle of the lathe, as regards height as well as angle, the angle gage, Fig. 6, is used

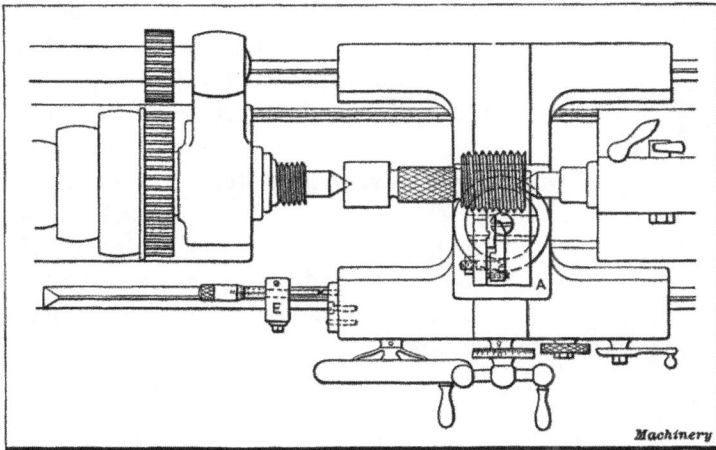

Fig. 13. Final Test of Pitch and Angle of Thread

in the same way as has been previously explained. When the fixture has been placed correctly in position, the screw plug is inspected by placing the gage first to the right and then to the left side of the thread angle. A strong magnifying glass is used with white paper underneath the gage, and any imperfection of the angle is easily detected, and can be corrected when lapping the last 0.0005 inch to size. If the test gage shows an opening either at the bottom or at the top, the fault is that the lap is worn and must be retapped, or it may be that too much emery has been used. If, for some reason or other, it is impossible to correct the screw plug within 0.0001 inch, when lapping, take a piece of hard wood, or flatten a piece of copper

wire, charge it with emery, and hand lap the high points of the
angle, while the screw plug is revolving slowly in the lathe.   In
this way, it is comparatively easy to overcome this trouble, but
great care must be taken to follow the thread properly with the
hand lap.

Testing the Lead of the Thread. — To find if a screw thread
has a perfect lead, the micrometer stop E, Fig. 13, is placed on
the left-hand side of the carriage.   The holder for this micrometer stop is shown separately in Fig. 14.   The construction of
the stop is very simple.   The micrometer head is an ordinary
one, as made for the trade by manufac-
turers of these instruments.   The holder
E is made similar to a C-clamp, with a
hole drilled and reamed to fit the micrometer head.   A slot is sawed through the
upper jaw, with a stop screw on the top,
which prevents the micrometer from being
clamped too hard in the holder, in which
case the thimble would not revolve freely.
Underneath this hole the holder is beveled
off, and a V-block F is held in position by
a screw G, entering from the side.   The
micrometer head is placed in the hole provided for it, with its graduations upward,

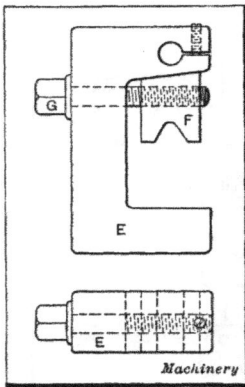

Fig. 14. Holder for
Micrometer Stop

and the screw G clamps the micrometer head and the holder
E at the same time.

When the lead of the screw plug is being tested, the carriage
is moved one inch along the thread.   It is understood that the
lead-screw of the lathe is not employed in this case, but one
depends upon the micrometer for measuring the correct lead of
the screw plug.   The master plug may, of course, also be placed
between the centers and comparison be made with the master
plug.   In this case, the micrometer serves as a comparator.   A
plate is screwed on the left-hand side of the carriage, provided
with a hardened stop against which the end of the micrometer screw bears.   It is evident that the carriage must not
be moved against the micrometer with too much force, but

simply brought up to barely touch the end of the micrometer screw.

Measuring Screw Thread Diameters. — It is always advisable when measuring screw thread diameters to measure them in the angle of the thread, in addition to testing the diameter over the top of the threads. Two methods for measuring angle diameters of screw threads are commonly employed. One of these methods, shown in Fig. 15, is used in connection with the Brown & Sharpe thread micrometer. In Fig. 16 is shown what is known as the three-wire method for obtaining angle diameters of screw threads.

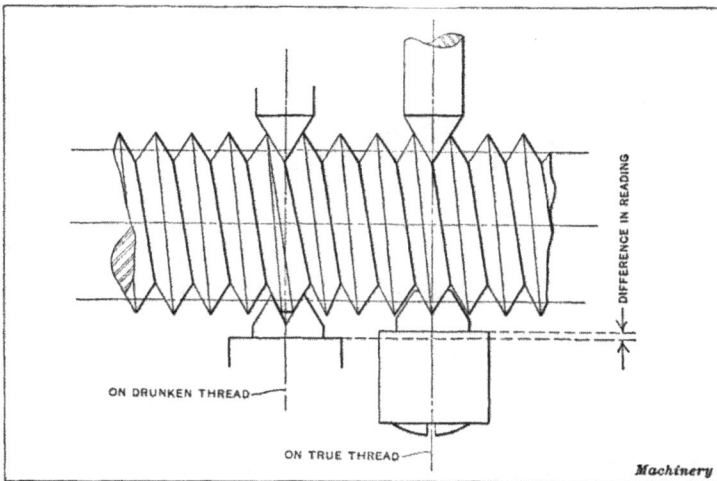

Fig. 15. The Anvil Type of Thread Micrometer gives Varying Measurements on True and Drunken Threads

These two systems may be used both for positive measurements and for comparisons. In Fig. 17 is shown a method of measuring with ball points inserted in the anvil and in the end of the measuring screw of the micrometer. This device is used for comparison only.

When the thread is correct in form, either the method shown in Fig. 15 or that shown in Fig. 16 will give equally good results for absolute measurements, and all three methods correct results for comparison; but if the thread should not be of the correct shape, but be "drunken," as indicated to the left in these three illustrations, then the anvil type of thread micrometer only shows this variation, while the three-wire method and the ball point micrometer would not indicate any error, provided the thread angle be correct. The reason for this is evident: the three-wire system measures the grooves cut by the thread tool, and as this is always set at the same depth and is unvarying in shape, the error, if any, would not be detected. The same condition is met with in the ball micrometer. The lower anvil point of a regular thread micrometer, however, since it spans the abnormal thread as shown in Fig. 15, instead of making contact

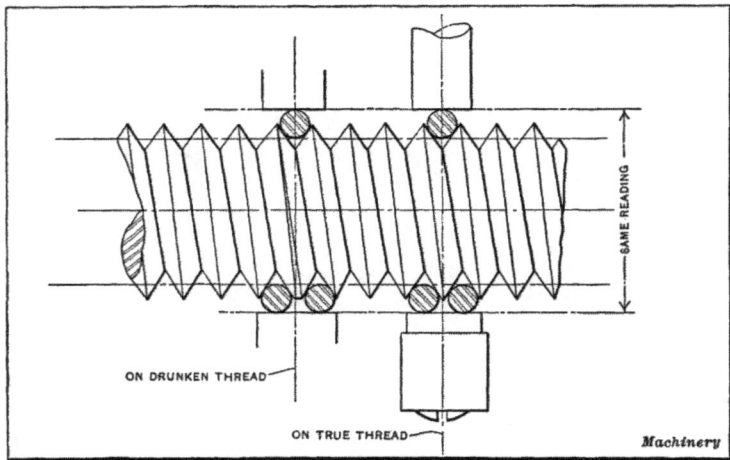

Fig. 16. The Three-wire Method of Thread Measurement makes no Distinction between True and Drunken Threads

with the sides of the adjacent threads, indicates the irregularity by giving an increased reading for the pitch diameter.

This condition, however, does not indicate that the three-wire method of measuring angle diameters is unreliable for ordinary use. With the methods used for accurate thread cutting in general, a drunken thread is seldom met with. If the thread is defective in a manner as shown to the left in the accompanying illustrations, then, as already mentioned, the Brown & Sharpe thread micrometer will indicate this defect, but the three-wire system nevertheless measures the pitch diameter correctly under all circumstances, as the principle of its use depends on the bear-

ing of the wire on the sides of the thread groove. The cases
when the thread is drunken when cut on well-made machine tools
are rare, and, therefore, the fact that the three-wire system does
not detect errors of this kind is not a very valid objection to its
use.

Three-wire Method of Measuring Threads. — When measuring threads by means of three wires and a micrometer, as illustrated in Fig. 16, wires of equal diameter should be used. In
order to determine if the pitch diameter of a screw is correct, first
ascertain (by the following rules) what the micrometer reading
should be, when using wires of a given size, and then compare the

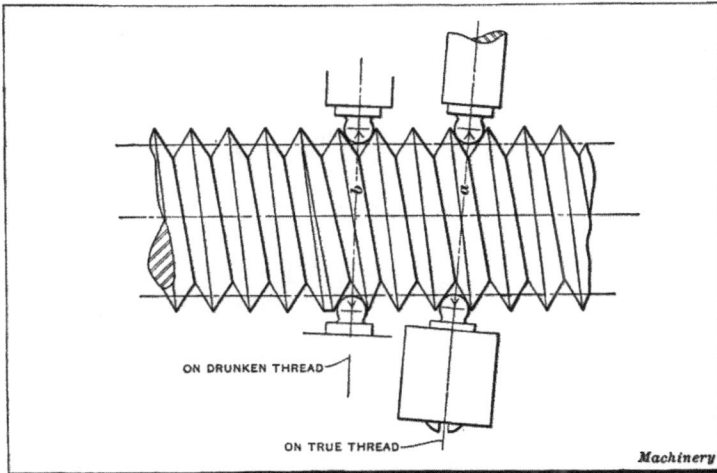

ON DRUNKEN THREAD

ON TRUE THREAD

Machinery

Fig. 17. The Only Difference in Measurement on True and Drunken
Threads made by the Ball-point Micrometer is due to the very
Slight Difference in Inclination — too Slight to be Appreciable

actual measurement over the wires with the calculated dimension. The micrometer reading for different standard threads
should be as follows:

For a U. S. Standard Thread: Divide the constant 1.5155 by
the number of threads per inch; subtract the quotient from the
standard outside diameter of the screw, and then add to the
difference three times the diameter of the wires: the result equals
the correct micrometer reading.

For a V-thread: Divide the constant 1.732 by the number of

threads per inch; subtract the quotient from the standard outside diameter, and then add to the difference three times the
diameter of the wires: the result equals the correct micrometer
reading.

For a Whitworth Thread: Divide the constant 1.6008 by the
number of threads per inch; subtract the quotient from the
standard outside diameter and then add to the difference 3.1657
times the diameter of the wires: the result equals the correct
micrometer reading.

Example. — Suppose a U. S. standard thread $1\frac{1}{2}$ inch in diameter and having 12 threads per inch is to be tested, and that
wires 0.070 inch diameter are to be used.

The micrometer reading for a screw of the correct size should
then equal $1\frac{1}{2} - 1.5155 \div 12 + 3 \times 0.070 = 1.5837$ inch.

CHAPTER V

THREADING CHASER TROUBLES AND REMEDIES

The troubles encountered by the toolmaker in keeping up a large assortment of thread chasers of various designs for cutting soft steel, tool steel, malleable iron, and especially brass, are many. There are also numerous methods advanced for overcoming the various troubles encountered, but they, like many of the theories regarding the manufacture and re-grinding of threading chasers, do not always prove satisfactory when put into practice. Owing to the large number of thread pitches used on different diameters there are, of course, many angles to contend with. This fact makes it impossible to control a cutter of the chaser class, as we do the single-pointed lathe tool for brass. This tool is ground with very little side clearance and with usually no top rake. We know that if the brass threading tool is not made in this manner, it is likely to run ahead of the feed, and produce a rough and uneven finish. This is the trouble which is also encountered with threading chasers held in a die-head, where they are required to make their own lead; they run ahead of the pitch, sometimes, to such an extent as to cut the thread entirely off the work. In other cases where the chasers are more correctly ground, they leave the rear side of the thread rough, which indicates that the die is leading ahead too fast. Sometimes just the opposite condition exists, that is, the front side of the thread is rough; this is caused by the die not leading fast enough.

The condition of the thread produced on both the front and rear sides should always be noted, as this is a guide which can be used to correct the trouble. Where it is necessary for the thread to fit sufficiently close to make a water-tight joint, it is desirable that the threads should be correctly formed. The ideal way to govern the leading of the die, is to grind more or less relief in the throat of the die, as the case demands, as indicated at A, Fig. 1.

The more relief given to the die, the faster it will lead, and the less relief the slower it will lead.   Unfortunately it is seldom possible to have a very long throat in the dies, especially on small brass work, where it is necessary to cut the thread up to a shoulder.   This condition makes it impossible to entirely control the free cutting of the chasers by the amount of relief in the throat.

Threading Die Chasers for Cutting Close to a Shoulder. — Where a set of chasers is to be used for cutting up close to a shoulder, instead of grinding the throat of the dies back, as shown at *A*, Fig. 1, the first two or three teeth in the chasers are lapped

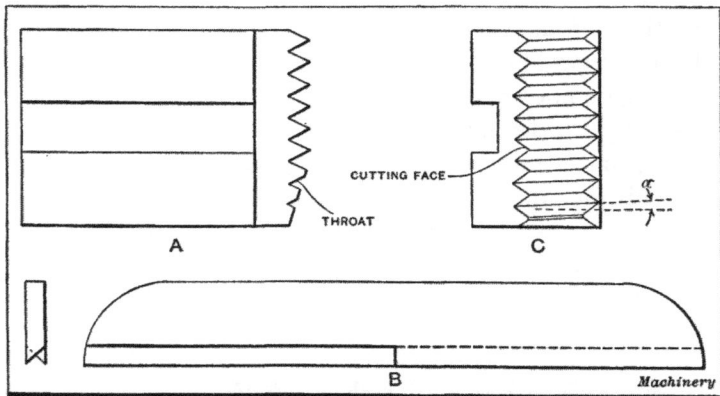

Fig. 1.   Illustrating how the Throat of a Chaser is Relieved — Tool
for Lapping the Sides of the Thread

with a strip made of sheet brass.   This strip (shown at *B*) is charged with carborundum and oil, and is moved across the first two or three threads by hand.   It should be held at the angle required to give the thread more or less side clearance as desired. The side of the thread to lap depends upon whether it is desired to make the die lead faster or slower.   If the angle of the thread α (shown at *C*) is reduced, the dies will lead slower.   The working edge of the lap should be made with an angle of about 50 degrees, so that it will be possible to get to the bottom of a 60-degree thread, and touch only the side desired.   When lapping a set of chasers in this manner, they should be tried out occasionally on a piece of work to see whether they are producing the desired results. A very slight change on the first two threads will generally be sufficient, leaving the rest of the cutting edges perfect, and free cutting.

The chasers should be sharpened frequently by taking the lightest possible grinding cut. The chasers should not be allowed to become so dull that it will be necessary for the grinder to remove so much material that the previous corrections on the chasers by lapping are entirely removed. Chasers are sometimes corrected with a three-cornered file, by filing the threads clear across the chasers instead of lapping the first two or three threads. Filing can be used for making the corrections, but unless the file is handled with a certain amount of dexterity, the results are not always gratifying. Frequently a chaser in a set is not so sharp as the opposite one, which results in the stock being forced over the sharper chasers. This only aggravates the condition, and makes it more difficult to find which chaser is producing poor results. Filing the chasers also causes them to burnish the thread instead of cutting, which puts an excessive strain on the die-heads, and when of the opening variety they frequently refuse to trip. Excessive heating of the work being threaded is also caused by the friction, which results in unequal expansion, and makes it difficult to assemble the parts, if they are to fit at all closely.

Angle of Top Rake for Threading Die Chasers. — The angle of the cutting face or top rake of the chaser is another point about which there is a considerable difference of opinion. In deciding on this angle, the material to be cut is the governing feature. When cutting machine steel, tool steel, or yellow brass, considerable top rake is necessary, as shown at *A*, Fig. 2. The idea of giving the top rake to the chaser is to produce a curling chip, so that the metal can be removed without undue friction. On cast brass the conditions are quite different, according to the alloy used. In most cases it is advisable to grind the cutting face radial as shown at *B*.

A point which has caused considerable confusion in the minds of mechanics is the constantly changing appearance of the cutting face of the chasers, after they have been resharpened a number of times.   Referring to C, Fig. 2, it will be seen that in order to maintain the top rake, the angle, if taken with reference to the chaser itself, and not with reference to its position in the die-head and the center of the work, seems to be greater as the chaser is ground away.   This may be the reason that a difference of opinion exists regarding the grinding of the chasers.   It is important, however, that the cutting face of the chaser should always bear the same relation to the work, and for this reason it is desirable to have a templet to grind the chaser to.   The templet shown at D has been found very useful for grinding thread

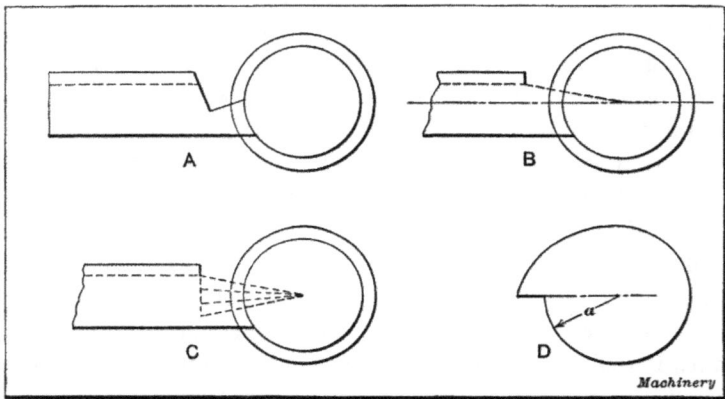

Fig. 2.   Method of Grinding Threading Chasers for Cutting Different Materials — Templet for Gaging Cutting Face

chasers for cast brass.   This templet is made of thin sheet steel and is held in the bottom of the thread.   The radius a is made one-half the diameter of the work on which the chasers are to be used.   The projection on the templet is ground to a radial line. By holding this templet in the bottom of the thread in the chaser, and grinding the face to the projection on the gage, the cutting edge will always be in alignment with the center of the work.

Position of Chaser in Die-head. — Experience has shown that chasers which gave satisfactory results in one die-head did not work at all when put in another die-head.   However, it was found after considerable experimenting, that the relief in the throat of the chaser should be controlled very closely; in fact, a difference of from 0.002 to 0.003 inch more throat relief would sometimes cause the thread to strip badly. This led to an investigation, and a comparison of releasing die-heads revealed a wide variation in the width of the slots. This accounted for the chasers working correctly in one die-head and poorly in another, as the change in the width of the slots would allow the chasers to tip more or less, thereby changing the throat relief.

In order to make a chart showing the exact position of the chaser in the throat grinding machine, so that it could be duplicated and used in any head when ground, it was found necessary to cut out the slots in the die-heads and insert hardened and ground tool-steel blocks in such a manner that they could be

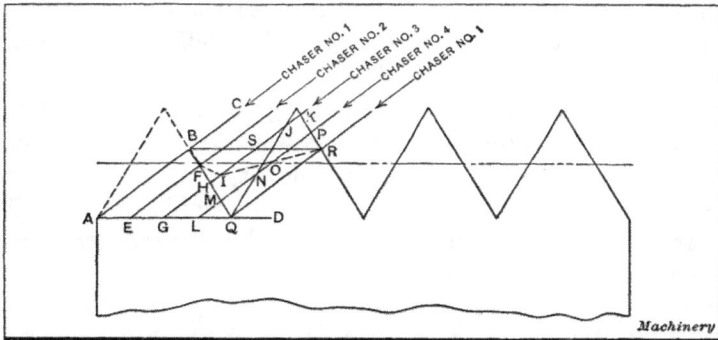

Fig. 3. Diagram Showing Variation in Length of Cutting Edge on Different Die Chasers of a Set

removed when worn and new ones put in their place. This permits keeping the slots in the die-head in good condition, and also facilitates making the chasers interchangeable. When a set of chasers for diameters of, say, 1 inch and 14 pitch, for example, is found by experiment to be correctly ground, the position they occupy in the throat grinding machine is marked on the chart so that it is possible to duplicate the grinding. Chasers which are made with a slight taper, say about $\frac{1}{4}$ inch to the foot, have been found to work much better on some jobs on account of good clearance conditions. It has been found that five times as many pieces can be threaded with such dies before resharpening, as can be done with ordinary straight ones.

Grinding Threading Die Chasers. — In the use of threading dies which must be ground at an abrupt angle to allow a full thread to be cut up to a shoulder, or to a neck, as in a set-screw, the following is a common procedure. When resharpening, the dies, or chasers, as the case may be, are put in a fixture which is so arranged that the angle of the cutting edge relative to the center-line of the die is the same on all the chasers, and also so adjusted that all the edges are ground to the same depth. When the die is used, however, it will often be found that in spite of all precautions for getting the chasers similarly ground, one or two of them will be doing most of the work, and will accordingly become dull while the others are in good condition. If the die is put back into the fixture and ground back an equal distance, on all chasers, using the same grinding angle as before, this will often result in an improvement. Sometimes a second grinding is necessary before the die will work properly.

This matter had been the cause of a great deal of annoyance, so that an investigation of the causes was made, which resulted in the following solution of the difficulty: By grinding the chasers in accordance with this method, the trouble entirely disappeared, and it became possible to grind the chasers so that each one always did the proper amount of work.

In Fig. 3, the line AB represents the cutting edge of a chaser. This chaser is ground at an angle DAC, with the center-line, which angle will be called the grinding angle, and this angle is selected at random, as is usually the case, being in this instance rather abrupt, for cutting a full thread close to a shoulder. The total cutting edge on chaser No. 1 is AB. The line EF represents the total cutting edge on chaser No. 2, counting the chasers in a left-hand direction on a right-hand thread, and on a four-chaser die. Lines GH and JK are the cutting edges of chaser No. 3, and LM and NP the cutting edges of chaser No. 4, while QR would be the cutting edge of chaser No. 1 again. In this case, chaser No. 3 is the first one to have the cutting edge divided into two or more parts. To get the total length of this, make HI equal to JK, then GI represents this total length. In the same way, LO is the total length of the cutting edge of chaser No. 4, MO being equal to NP. It will readily be seen that these cutting edges are not of the same length, and the discrepancy is made more graphic by connecting the points *BFIOR*; and also the points *BR* with a straight line.   The line *BFIOR* will be called the indicating line.   The distance *IS* is the amount that the cutting edge of chaser No. 1 is longer than that of chaser No. 3.

In service, a die having its chasers ground in this manner, will act as follows: Chaser No. 1 will take a broad, thin chip, and will crowd the work over on the opposite chaser, No. 3, which will be forced to carry a very thick chip, and if there is considerable difference in the length of cutting edges, this action seems to be all out of proportion to this difference.   In fact, in this case, with the cutting edge of No. 3 only about 25 per cent less in length

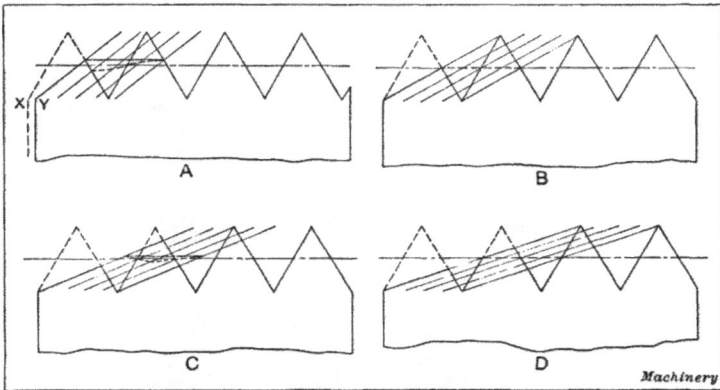

Fig. 4. Illustrations Showing how Die Chasers should be Ground so that Each One in a Set will do an Equal Share of the Work

than No. 1, it probably would carry a chip three or four times as thick.   By grinding the chaser back a distance *XY* (sketch *A*, Fig. 4), with the cutting edge at the same angle as in Fig. 3, it will be noted that the indicating line shows a much smaller variation in the length of cutting edges, those of No. 2 and No. 3 being approximately equal, and No. 1 and No. 4 also being nearly the same.   The chasers will now show a marked improvement in operation.   If they are ground back a little further the old trouble will reappear.

The diagram *B* (Fig. 4) illustrates a chaser which is ground at an angle of 30 degrees with the center-line; that is, the grinding line connects the bottom of one thread with the top of the next thread behind. In this case it will be found, by measurement, or by reference to geometry, that the indicating line becomes straight, parallel to the center-line, and midway between the top and bottom of the threads. This indicates that by grinding the chasers at this angle, the cutting edges become equal, and practice proved that each chaser will do its proper share of the work.

In diagram C, the grinding line connects the bottom of one thread with a point between the tops of the second and third threads behind, which makes the grinding angle less than 30 degrees and greater than 19 degrees. Here the indicating line deviates again from a straight line, but this deviation is very

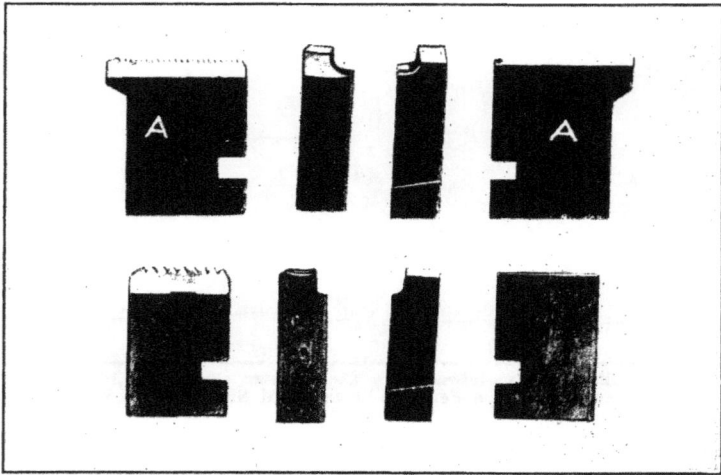

Fig. 5. Threading Die Chasers used for Brass Work

much less than in the first example. Diagram D shows a grinding angle such that the grinding line connects the bottom of one thread to the top of the third thread behind. Considering the pitch as the unit, this angle will be seen to be such that its tangent $= \dfrac{0.866}{2.5} = 0.34641$. This makes the angle a trifle over 19 degrees. It will readily be seen by comparing similar sides of similar triangles, that the cutting edges on all four chasers have the same length, and the indicating line is straight, as at B.

In general, in order to make the cutting edges of all the chasers equal, where the thread angle is 60 degrees, it is necessary to make the grinding angle such that its tangent will be equal to 0.866 ÷ (a whole number plus 0.5), or, in other words, so that the grinding line will always be parallel to a line connecting the point at the root between two threads, with the top of some other thread.   In a 60-degree thread, these angles will be found to be 60, 30, 19, 14, 11 degrees, etc.   When ground at any angle less than 14 degrees, that is, with the chip distributed over four threads or more, the difference in the total lengths of cutting edges becomes negligible,

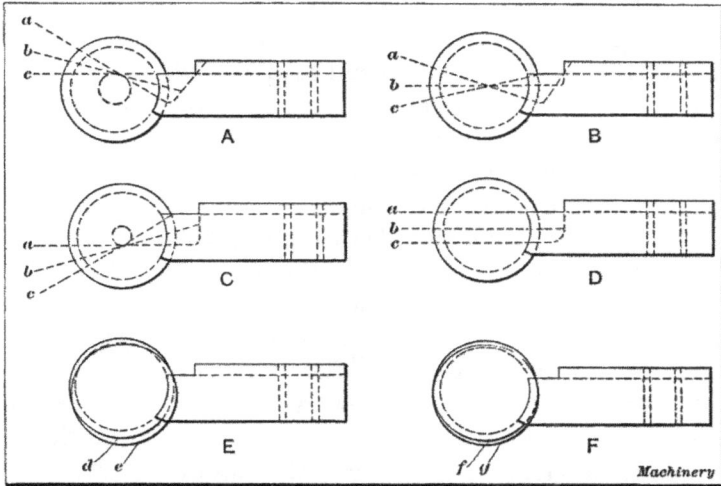

Fig. 6.   Diagrams Illustrating Points on the Grinding of Chasers for Brass

and no attention need be paid to the angle other than to see that all chasers have the same angle, but when the angle is abrupt, so that two or three threads do all the work, an observation of the foregoing principles for securing equal lengths of cutting edges, will solve a surprising number of die problems.

Grinding Die Chasers for Threading Brass. — When brass is reasonably uniform in hardness, the difficulty of grinding threading dies to suit the work is not so pronounced, but when all kinds of brass (scrap as well as new brass) are used, considerable experience is required to sharpen the thread cutting chasers so that they will give satisfaction. Definite rules that will cover all cases cannot be given, but a few general principles applying to the grinding of these tools may be of value. That difficulties will arise on account of ununiformity in the texture of the brass to be worked upon is, of course, evident. Another difficulty to be contended with is the wear of the die-heads. When the die-heads are new, the chasers are held reasonably solid, but when the heads wear, the chasers become loose and must be ground so that they will not dig into the brass and produce a torn or imperfect thread. The following is descriptive of the practice of grinding chasers in a shop where a great deal of this work is done. All chasers used in this shop are made from Jessop steel. The threads, after being cut and relieved, are hardened and then ground.

In making brass chasers it is absolutely necessary that the threads be relieved; otherwise clogging will result. In the class of work just referred to, most of the threads are run up close to a shoulder, and, consequently, a sharp chamfer must be ground on the chasers. It is not advisable to grind the chamfer any oftener than is necessary, because each grinding of the chamfer will shorten the life of the chasers in those cases where the die cuts close to a shoulder. When chasers show a tendency to break off at the chamfer or to get dull quickly at this place, they may be made with an extension as shown at A in Fig. 5. When made in this way the chasers are usually worn out by the time the extensions are ground off flush with the face of the body.

Diagram A, Fig. 6, shows a chaser ground to angles advantageous for very hard brass. The angle is such that the face of the cutting edge is $\frac{1}{8}$ inch ahead of the center at the successive grindings. At C the chaser is ground with a negative rake, the cutting edge being $\frac{1}{16}$ inch below center. This form is used occasionally for certain classes of work, such as very soft "greasy" brass. The chaser shown at B is ground in a way to give the best satisfaction for all-around work. The cutting edge is ground so that a line passing through the face also passes through the center of the work. On the other hand, diagram D shows a very poor way to grind a chaser whether for brass or steel, as the angle of the cutting point changes with each successive grinding, and while such a chaser may cut satisfactorily on some work when nearly new, it is not likely to do so after one or two grindings.

In grinding the chamfer on a brass cutting chaser, a wheel about the size of, or very slightly larger than, the piece which the chaser is intended to cut, should be used. The relation of the wheel to the chaser when grinding the chamfer is indicated at E and F. In sketch F, the circle f represents the stock to be cut, and g the emery wheel used for grinding the relief of the chamfer.

Fig. 7. Grinding the Chamfer on a Chaser

As will be seen, the center of the emery wheel is a trifle below the center of the stock, and somewhat toward the right. At E is shown the relation when grinding a chaser for hard brass. Extreme care must be taken, in grinding, not to draw the temper of the cutting edge.

As a general rule, except for the very hardest brass, there should be as little relief as possible back of the cutting edge of the chamfer, because this part of the die steadies the chasers when starting the thread.   It is of advantage when grinding new chasers to set the chaser well over to the left of the wheel, using a fixture such as shown in Fig. 7, and just touch the edge of the chamfer farthest away from the cutting edge, with the wheel; then gradually move the wheel over until there is a slight relief all the way up to the cutting edge.   A fixture or jig used for grinding the faces of chasers is shown in Fig. 8.

Alundum wheels of rather fine grade and known as elastic

Fig. 8.   Grinding the Face or Cutting Edge of a Chaser

bond, have proved satisfactory for grinding chasers but no matter what wheel is used there is a fin left on the cutting edges of the tool, which is a source of trouble if not removed.   Lapping in various ways was tried for removing this fin, but the most satisfactory and quickest way, is to use a very fine three-cornered file and just draw it lightly down each V-groove of the thread at an angle of about 45 degrees, not pressing hard enough to more than take away the fin and slightly dull the cutting edge.   An oilstone should then be run lightly over the edge of the chamfer to dull it just enough, so that it is not entirely sharp.   The chaser threads are then dipped in a little flour of emery and oil, and brought up against a wire brush, as shown to the left in Fig. 9. They are then turned over and brought up against the wire brush as shown by the view to the right.   This should be done very lightly, and has the effect of smoothing the edges the right amount.   A little practice soon enables one to determine how much of this treatment is required to give the best results.   The wire brush is 6 inches in diameter, made of fine wire, and runs 3400 revolutions per minute.

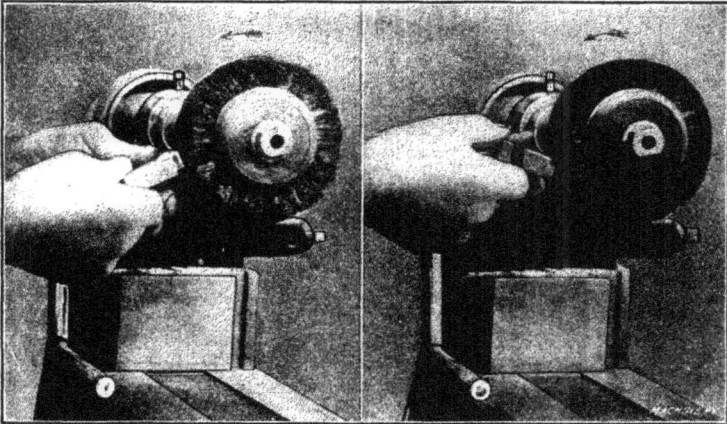

Fig. 9.   Revolving Wire Brushes which smooth the Cutting Edges of Chasers

As a rule, a brass cutting chaser will chatter when too sharp. When the thread cut on the work is torn, it is evident that the chasers are dull, or that they feed too fast.   Burrs in the thread tend to have the same effect as increasing the pitch in many cases. Chasers generally tend to feed too fast rather than too slow, and the trouble can sometimes be remedied by the removal of burrs or dulling of the cutting edges, provided, of course, that the teeth are already ground to the right cutting angle.   When a set of chasers feed too slow the cause is usually that there is not enough clearance back of the cutting edges of the chamfer.   Whether chasers feed too fast or too slow is easily determined by examining the thread cut.   If the side of the thread next to the turret is smooth, and the side next to the chuck is torn or ragged, the chasers feed too fast, and *vice versa*.

Making Accurate Thread Chasers. — To make an accurate chaser of the type shown at *C*, Fig. 10, is quite simple when properly understood. The first thing is to make the hob, which is shown at *A*. This requires great care because the accuracy of all the tools of that particular pitch depends upon the hob. It is the practice of one concern to make the hobs all

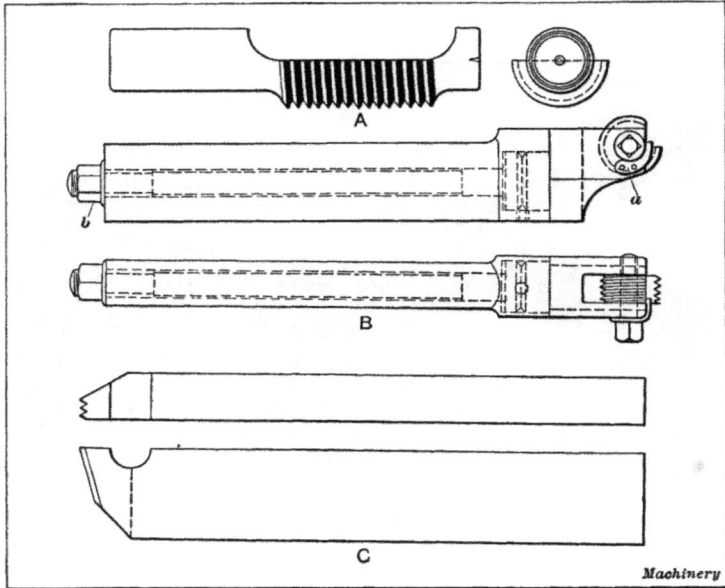

Fig. 10. (A) Hob. (B) Circular Chaser. (C) Chaser made by Hob shown at A

one inch in diameter, this size having been adopted so that the angles in the accompanying table could be determined and tabulated. To accurately cut the hob, the tool shown at *B* is utilized. This consists of a small circular thread chaser held in the body of the tool, the forward part of which is made separate from the shank so that it can be swiveled to suit the angle of the thread on the hob. A small piece of steel, *a*, serves as a gage for the cutting face of the circular chaser, so that it can be sharpened and re-set in the holder without disturbing the body

of the tool.   The nut *b*, on the end of the holder, serves to hold
the forward part of the tool securely.   After the hob is threaded,
it is milled out (as shown at *A*) to its center-line and then
hardened.   The object in milling it in this manner is that it
can easily be sharpened by grinding across the face, and this
face is also utilized when setting the hob to its proper angle in
the milling machine.

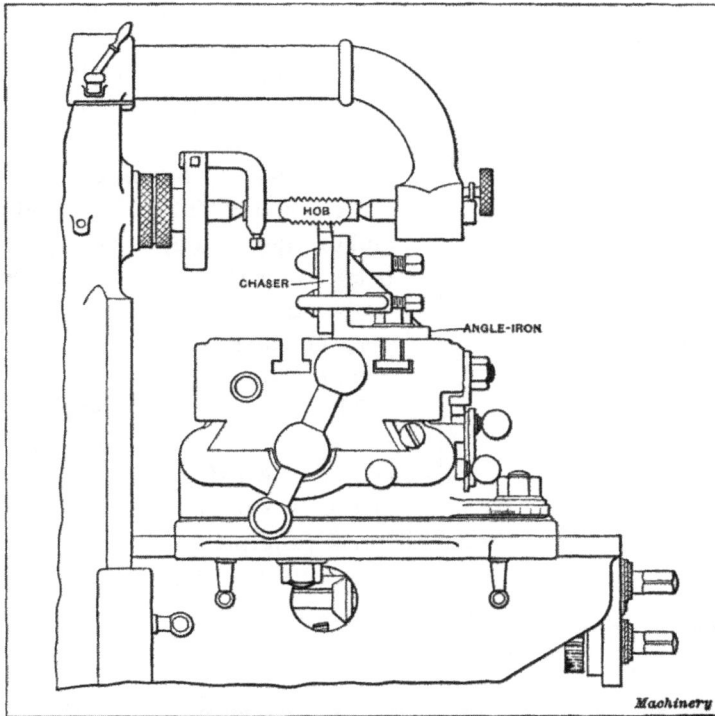

Fig. 11.   Milling Machine arranged for Hobbing Chasers

The sharpening of the hob is accomplished with the special
fixture similar to the one shown in Fig. 4, Chapter III.   This
fixture is simple in design, and is made for use on a surface
grinder, where it is located so that its centers are at right angles
to the grinding wheel.   The most essential point in grinding
a hob of this description is to always grind the cutting face
radially; in other words, the lower edge of the grinding wheel

must be in line with the center of the hob.   To set the grinding wheel, lever *a* is employed.   This lever has its fulcrum on the block *d* while its other end extends to the forward block upon which are graduated a few lines, about 0.05 inch apart, each division equaling a movement of 0.001 inch at the ball *b*.   To set the grinding wheel, the rear center *c* is removed from the block *d*, and the grinding wheel, at rest, is brought down onto the ball *b*.   The table of the grinder is run to and fro by hand so that the wheel will pass over the ball; when it forces down the lever so that it registers at zero, this denotes that the lower edge of the wheel is in alignment with the centers of the fixture.   The center *c* is then put back in position and the hob, held by a dog, is placed between the centers and sharpened. At no time during grinding is the perpendicular adjustment of the wheel altered.

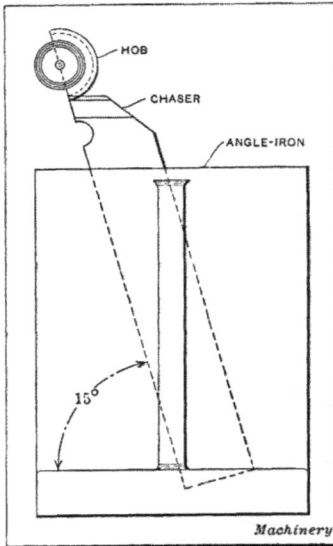

Fig. 12.   **Hob and Chaser in Position for Hobbing**

The hob being completed, the next step is to use it in making the chaser shown at *C* in Fig. 10.   This chaser is made of tool steel, hand forged, and planed on all sides.   It has a cutting clearance of 15 degrees and is placed against an angle iron which, in turn, is held on a milling machine table.   The hob is held between the centers of the machine spindle and the overhanging arm, as shown in Fig. 11, and when the cutting edge of the hob is accurately located, the spindle is locked in position by means of a wooden wedge which (in the case of a belt-driven machine) is tapped in between the cone and the frame of the machine. The cutting face of the hob and the body of the chaser should be in the same plane and at an angle of 15 degrees with the table of the machine, as shown in Fig. 12.   The most essential point in setting up the machine for this job is to get the angle iron located on the table of the machine at the proper angle for the threads to be shaped on the chaser, as a chaser made for use on a ½-inch tap will not work properly on a 2-inch tap of the same pitch, because the angle of the thread is greater on the former than on the latter. The milling machine table must also be

Angular Position of Machine Table and Work-holding Fixture, for Hobbing Chasers

Threads per inch.	Angle B in degrees and minutes.																Angle A	
	Deg.	Min.	Deg.	Min.	Deg.	Min.	Deg.	Min.	Deg.	Min.	Deg.	Min.	Deg.	Min.	Deg.	Min.	Deg.	Min.
8	..	...	..	...	3	02	2	17	1	49	1	31	1	18	1	08	2	17
10	..	...	3	38	2	26	1	49	1	27	1	13	1	02	..	55	1	49
12	..	...	3	02	2	01	1	31	1	13	1	00	..	52	..	46	1	31
14	..	...	2	36	1	44	1	18	1	01	..	52	..	44	..	39	1	18
16	..	...	2	17	1	31	1	08	..	55	..	46	..	39	..	34	1	08
18	..	...	2	02	1	21	1	00	..	49	..	40	..	35	..	30	1	00
20	3	38	1	49	1	13	..	55	..	44	..	37	..	31	..	27	..	55
22	3	19	1	39	1	06	..	50	..	40	..	33	..	28	..	25	..	50
24	3	02	1	31	1	00	..	46	..	37	..	30	..	26	..	23	..	46
26	2	48	1	24	..	56	..	42	..	34	..	28	..	24	..	21	..	42
28	2	36	1	18	..	52	..	39	..	31	..	26	..	22	..	19	..	39
30	2	27	1	13	..	49	..	37	..	29	..	24	..	21	..	18	..	37
32	2	17	1	08	..	46	..	34	..	27	..	23	..	19	..	17	..	34
36	2	02	1	00	..	40	..	30	..	24	..	20	..	17	..	15	..	30
40	1	49	..	55	..	37	..	27	..	22	..	18	..	16	..	14	..	27
48	1	31	..	46	..	30	..	23	..	18	..	15	..	13	..	12	..	23
56	1	18	..	39	..	26	..	19	..	16	..	13	..	11	..	10	..	19
64	1	08	..	34	..	23	..	17	..	14	..	12	..	10	..	08	..	17
80	..	55	..	27	..	18	..	14	..	11	..	09	..	08	..	07	..	14
100	..	44	..	22	..	15	..	11	..	09	..	07	..	06	..	05	..	11
	¼″		½″		¾″		1″		1¼″		1½″		1¾″		2″			

Diameter on which chaser is to be used.

NOTE. — All hobs to be 1 inch in diameter right-hand thread. Clearance on chasers, 15 degrees.

swiveled around to the proper angle of the thread on the hob, as the longitudinal movement of the table must correspond to the thread angle. These angles may be obtained from the table previously referred to, headed, "Angular Position of Machine Table and Work-holding Fixture for Hobbing Chasers." The plan view above this table will serve to illustrate its use. This is a plan of the milling machine table, showing it swiveled around, and also the angle iron set in the proper position. As will be seen, the table sets at an angle A, which is given in degrees and minutes in the right-hand column of the table, while the angle iron which forms the work-holding fixture is located at an angle B with the edge of the table. This is the proper angle for the threads on the chaser. Should it be desired to make a left-hand thread chaser, the angle iron would be placed at the same angle given in the table, but in the opposite direction.

As an example, we will suppose that it is desired to make a chaser that is to be used in making taps $\frac{1}{2}$ inch in diameter having 26 threads per inch. First find 26 in the "threads per inch" column; then by following along the line to the last column, the angle at which the milling machine table is to be set, or the angle A, is obtained. This angle equals o degree and 42 minutes. On the lower edge of the table are the diameters on which the chasers are to be used, and as the diameter in this case is $\frac{1}{2}$ inch, we follow up that column to the 26 threads per inch line, where we obtain the angle at which the angle iron is to be located on the table of the machine, or angle B, which is 1 degree and 24 minutes. The machine being properly set, it is a small matter to shape the thread by moving the table to and fro, gradually feeding it upward until a perfect thread is obtained on the chaser. It is advisable to keep the hob well lubricated when cutting to insure a smooth thread on the chaser. A very good lubricant for this purpose is a mixture of one-half turpentine with one-half good lard oil. This will also be found an excellent lubricant for general thread cutting in the lathe.

CHAPTER VI

GENERAL TOOLMAKING OPERATIONS

Originating a Straightedge. — An accurate method of originating a straightedge is based on the principle that three straightedges cannot fit together, interchangeably, unless the edges are straight or plane surfaces; therefore, when employing this method, it is necessary to make three straightedges in order to secure accurate results. The general method of procedure is as follows: Three blanks are first machined as accurately as possible and these should be numbered 1, 2 and 3. That one of the three blanks which is believed to be the most accurate is then selected as a trial straightedge and the other two are fitted to it; for instance, if No. 1 is to be the first trial straightedge, Nos. 2 and 3 are fitted to it. This fitting should be done so accurately that no light can be seen between the straightedges when No. 1 is in contact either with No. 2 or No. 3. The accuracy of Nos. 2 and 3 will now depend upon the accuracy of No. 1. If we assume that No. 1 is slightly concave, as shown exaggerated at A, Fig. 1, then Nos. 2 and 3 will be convex; hence, the next step in this operation is to place Nos. 2 and 3 together to determine if such an error exists. As the diagram B indicates, the inaccuracy will then appear double. One of these blanks, say No. 2, is next corrected and made as near straight as possible. When making this correction, one should be guided by the error shown by test B. No. 2 is then used as the trial straightedge and Nos. 1 and 3 are fitted to it. The entire operation is then repeated; that is, blanks 3 and 1, after being fitted to No. 2, are placed together to observe whatever error may exist and then No. 3 is changed as indicated by the test. No. 3 is then used as the trial straightedge and Nos. 1 and 2 are fitted to it. The latter are then placed together; No. 1 is corrected and again used as a trial straightedge, to which Nos.

2 and 3 are fitted. By repeating this operation, the original error will be gradually eliminated and a straight surface originated. Of course, if an accurate surface plate or other plain surface is available, the method just described will be unnecessary as the straightedge could be fitted directly to the surface plate.

The method of finishing the edges depends not only upon the form of the straightedge, but, to some extent, upon the material of which it is made and, in the case of steel, whether the edges are hardened or not. Large straightedges, which to secure greater rigidity are often made of cast iron, can be finished

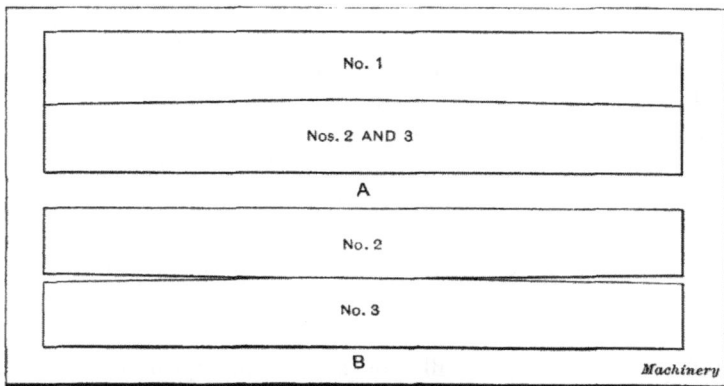

Fig. 1. (A) Straightedges Nos. 2 and 3 fitted to No. 1. (B) The Inaccuracy copied from No. 1 is revealed when Nos. 2 and 3 are placed together

after planing by carefully scraping the edges. Soft steel straightedges having plane surfaces or edges can be finished by scraping and lapping, whereas, if the edges are hardened, the high "spots" or areas can be reduced by stoning or lapping. The edge should preferably be ground on a surface grinder; then, any slight errors which may exist can be lapped down by using a cast-iron lap similar to the one shown in Fig. 2. As will be seen, this lap is simply an L-shaped piece of cast iron which is charged along the under surface A. This charged surface is at right-angles to the side of the lap which is held against the side of the straightedge when in use, to prevent the lap from being canted.

Originating a Surface Plate. — The method of originating straightedges by fitting three together until any two are a perfect fit (as near as can be determined by a practical test) can also be applied to the making of surface plates. Obviously, if only two plates are put together, they may not have true plane surfaces even though they show a good bearing when tested. This is because the inaccuracy in one place will often be concealed by corresponding inaccuracies in the other. Therefore, to secure accurate results, three plates should be scraped in together, these being numbered 1, 2 and 3. First, fit Nos. 3 and 2 to No. 1. When this has been done, Nos. 2 and 3 are, practically speaking, duplicates. The second step is to fit Nos. 2 and 3 together by scraping about as much from one plate as from the other in order to reduce any error which may have been copied from No. 1; third, fit No. 1 to No. 2; fourth, fit No. 3 to No. 1; fifth, fit No. 2 to No. 3 by scraping as much from one plate as the other. Continue this series of operations carefully until plate No. 1 will fit No. 2 and No. 3, and No. 2 will fit No. 1 and No. 3. Having originated three plates in this way, one can be laid aside to be used as a master-plate for testing the others which

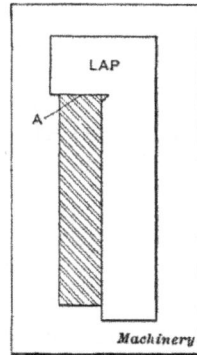

Fig. 2. Lap for Straightedges

are employed in active service. The time required for making surface plates by this method naturally will depend largely upon the accuracy of the machined surfaces. The judgment exercised in scraping down the bearing surfaces to secure an even distribution of the bearing marks also plays an important part.

Making Surface Plates by Lapping. — To originate surface plates by scraping three together, as described in the foregoing, usually requires considerable time. A more rapid method, which is sometimes employed, is as follows: First rough off the surface and the feet, taking off at least three-eighths inch to make sure that all the surface iron is removed. This will prevent warping. Then drill and tap the holes for the handles. The plates should then be allowed to season for at least two months to allow the internal strains to permanently adjust themselves. To finish the plates, all that is necessary is a small amount of carborundum for lapping the plates together. Before starting to lap, stamp the plates No. 1, No. 2, and No. 3, respectively. Lay No. 2 on the bench, sprinkle some No. 36 carborundum over it, add a little coal oil, and lap No. 1 with it, using a circular motion, and constantly changing the relative position of the upper plate until something like a finished surface is obtained. Then lap No. 2 and No. 3 in the same manner, then No. 3 and No. 1, and then do the whole operation over again. By this time reasonably plane surfaces should have been obtained. Then repeat the procedure, using No. 60 carborundum grain. Then use No. 120, and follow this with No. 220 for the final finishing. The result will be three surface plates with true planes, a good finish, and a uniform bearing all over. If the lapping has been carefully done, these plates will lift each other by cohesion.

Some may think that this method of lapping plates together will so charge them with the abrasive material as to make them unfit to use with fine tools. However, when it is considered that cast-iron holes in accurate watch machinery are lapped to a bearing, and put to every-day use without wearing out the spindles or shafts that run in them any sooner than those used under the same conditions but not lapped, it is doubtful if surface plates can be loaded with abrasive material to such an extent as to make them unfit for use with fine tools. The author has used this process for making plates 12 inches square, and secured very accurate results.

Making Accurate Arbors. — An arbor, to be an efficient tool, should have a true cylindrical surface, and have center holes that are exactly coincident with the axis, round, true as to angle, and perfectly smooth. The most important of these requirements is the center hole, as it is easily deranged. An arbor can very easily be ruined in a few seconds by letting one of the center holes get dry and roughen. The shape or design of a center hole is a factor that must not be looked upon as un-important, although it is largely a matter of choice. The center shown at *A*, Fig. 3, is reamed to standard angle and the corner is rounded off at the mouth, while the center illustrated at *B* is

reamed to standard angle and the end of the arbor counterbored. The advantages claimed for the latter are, protection
of the center hole, and that it will stand considerable driving
without spoiling the center hole, but then, driving is rather a
barbaric practice. If by chance the arbor shown at *B* should
get caught by the lathe center near the center hole, the point
of the center is liable to raise a small burr, or chip out a small
piece at the edge of the center hole. With the center shown at
A, this is prevented by the center striking the curved edge of

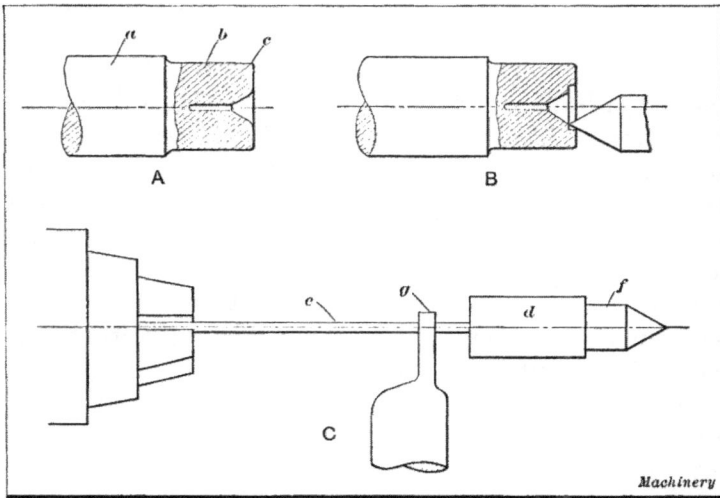

Fig. 3. Arbor Center Holes and Arbor Lap

the hole, and sliding into place. If the point strikes back of
the curve it will be too far from the center hole to do any damage.

When making arbors the stock is first cut to length, and then
carefully centered so that there will be a uniform amount of
stock removed from all sides of the piece. This is important
because of the difference of density caused by the rolling operations. After taking a roughing cut over the pieces, it is advisable to pack anneal them. This is to give any internal
strains which were caused by the stock being straightened cold
after being rolled, a chance to adjust themselves. After annealing they may be turned down, leaving the part *a* (Fig. 3)

large enough to allow for grinding. This allowance will depend upon the length and size of the arbor. The possible spring of the stock in hardening must also be considered. The part *b*, with the curved edge *c*, may be turned to size and polished. Hold one end of the arbor in a true universal chuck, and run the other end in a steadyrest. The curved edge of the center hole is now finished with a hand graver, and the end polished. Procure a drill the size of the teat of the center drill used, and drill the center hole to the depth of about $\frac{1}{2}$ inch. This is to make a reservoir for the storage of oil to lubricate the center when in use.

We are now ready to ream the center holes. This is a very important operation, and one that cannot be hurried, for much depends upon this part of the work. We will need a 60-degree center reamer, one having several flutes preferred. These flutes should be honed so that they have smooth, keen edges. Grip the center reamer in a chuck so that it runs absolutely true and block the lathe spindle. Now flood one center hole of the arbor with lard oil, and support the other end on the tailstock center which has been carefully lined up with the live spindle. Ream out the center hole by slowly turning the arbor by hand and tightening up on the tail center enough to make the reamer cut. Take very light cuts at a slow speed and use plenty of oil. Ream sufficiently to clean up the center hole from the center to the edge of the curved mouth. The arbors are now ready to be hardened. If the hardening is done with care, the majority of them will be fairly straight. Large arbors may be left hard, but arbors below $\frac{3}{4}$ inch should be drawn, so that they will not be so liable to break. The practice, at the present time, among manufacturers, is to harden arbors all over, but for extremely accurate work, an arbor having hardened ends and a soft body is generally considered superior, as there is less tendency of distortion from internal stresses. Hardened arbors should be "seasoned" before finish grinding to relieve internal stresses.

Lapping Arbor Center Holes. — An accurate center hole is the most important part of an arbor, and to get this, it should be carefully finished by lapping. Brass laps seem to work very well, and as the lap has to be of correct shape to get a true hole, it is advisable to make up a quantity of them, so that the temptation to continue using a lap which is worn out of shape will be removed. The lap should be gripped in a small chuck or holder, and run at a very high speed. A speed lathe with a lever feed is excellent. If the lathe is not provided with a lever feed, remove the tail-stock cap so that the spindle can be moved in and out very easily by taking hold of the tailstock screw. For the rough lapping use coarse emery. Drop a little oil in the center hole, and then sprinkle in a small amount of the emery. Support the other end of the arbor on a center in the tail-spindle and gently feed the arbor onto the revolving lap, removing it as soon as it touches the lap. Repeat the operation, and keep turning the arbor to neutralize any error that may exist in the alignment of the machine. It is important to do the lapping by instantaneous contacts with the lap to prevent the emery getting caught and "ringing" the center hole. It is much easier to keep the hole smooth, than to smooth it up after it is scored. Lap just enough to correct any distortion that may have been caused by the hardening. After rough lapping, change the lap, and with a finer grade of emery, re-lap the center holes to get them smoother, and to correct any errors that may have occurred in the roughing out. A new and perfectly true lap should be used for finishing the centers. As the abrasive toward the mouth of the center moves more rapidly than that farther in, and as centrifugal force tends to move it to the largest part of the hole, the lap will cut faster at this point. Because of this tendency, centers which are to be hardened and lapped are sometimes reamed to an angle (say 59 or $59\frac{1}{2}$ degrees) somewhat less than 60 degrees.

After the holes are lapped true and to the required angle they can be polished by the device shown at C, Fig. 3. Make a small brass cup d with a flexible steel shaft e about 4 inches long and $\frac{3}{32}$ inch in diameter. With sealing wax, imbed a small piece of oil-stone f in the cup. Then rig up a small rest with a short bearing at g, about $\frac{1}{2}$ inch back of the cup. The oilstone can be trued up with a diamond or piece of emery wheel, to the correct 60-degree angle. The object of the flexible shaft is to neutralize any chatter that may be caused by entering the lap in the center hole. This chatter, if not provided for, will cause the point of the oilstone to chip off. The rest is to keep the lap revolving in one plane. If this is not done, the lap revolving at a high speed is liable to go down through the shop if, by chance, it gets a little off center. Clean out all the center holes with gasoline to remove all emery, and insert the lap, using plenty of lard oil. Bear lightly on the lap to prevent it from getting dry and sticking, which would cause the point to break off. The oilstone will require truing up for every fifteen or twenty holes lapped, but considering the quality of work done, and the rapidity, it is a refinement well worth while. When the lapping is done, carefully test all the holes with a standard 60-degree plug, and mark all that will need to be lapped again to correct errors.

The last operation is to grind the cylindrical part. Because of the strains caused by the hardening, it is advisable to let the arbors season between the roughing and the finishing cuts. It will be found in some cases that they change shape after the roughing cut is taken. The arbor, if tapered, should have the number on the large end; the amount of taper is governed by the length and the use to which it is to be put. Arbors are usually tapered about 0.006 inch per foot.

The steel used for making arbors should not have too high a carbon content, as such steel is more liable to crack or spring out of shape when it is being hardened. A steel for arbors of medium size, having one per cent of carbon, will give good results.

Making an Internally-threaded Arbor. — A job which a toolmaker sometimes meets with is illustrated in Fig. 4 which shows an arbor having a threaded hole. This threaded hole should run perfectly true with the outside taper, and with the end E. The hole is to be sized with a tap so as to be of a standard dimension. The method of making this arbor will be described briefly in detail.

Cut off a piece of stock to length, allowing for facing. Face both ends in a chuck to nearly the finished length; then center both ends carefully. Rough turn the arbor, and turn the end E

true and smooth to a size slightly over its finished size. Place the steadyrest in position so as to guide the arbor at S, S. A dog should be strapped to the faceplate so as to hold the arbor tightly against the spindle center. Remove the tailstock center from the work and, after securing the steadyrest, run the tailstock center up to its place again and examine the center in the end E of the arbor to make sure that the steadyrest has not sprung the arbor out of line. If everything seems all right, push

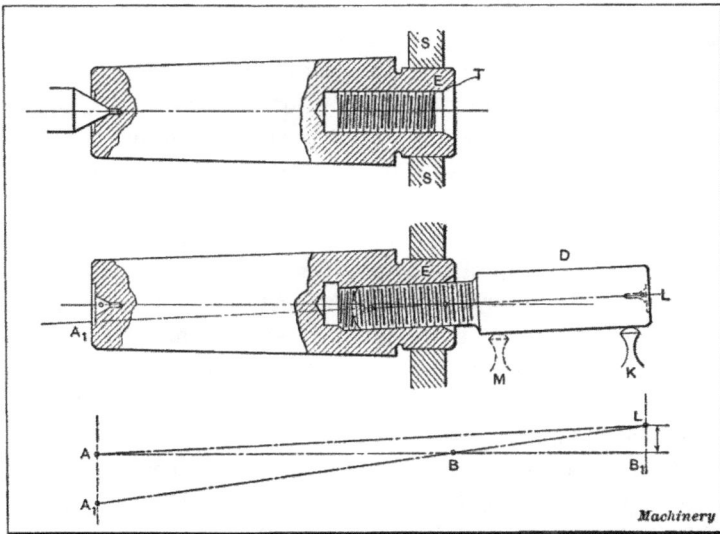

Fig. 4. Internally-threaded Arbor

back the tailstock, and proceed to drill and bore the hole. To get this perfectly true, it should be finished with very light cuts at a slow speed. The size of the hole should be slightly larger than the bottom diameter of the thread of the tap with which the hole is to be sized. It is difficult to fit a screw to a hole that has a full V-thread. After boring the hole, recess its end, as shown, enlarging it to a little more than the full diameter of the thread to be cut. This makes a clearance space for the thread tool when cutting the thread. Then enlarge the end E, and bore the taper T carefully so that it can be used as a center later on.

The thread should be cut carefully, making sure that the thread tool is set so as to cut a symmetrical thread. During the last few cuts the work should revolve slowly, and light cuts should be taken. Be sure the tool is hard and keen. After cutting the thread nearly to size, finish the hole with the taps, first using a taper tap, next a plug, and last a bottoming tap. The usual chance for error in a job of this kind lies in the tapping of the hole. If not carefully done, the tap gets started out of true, and when finished, the thread in the hole is out of line with the center. This error is shown exaggerated at D in Fig. 4. It is essential to test the arbor to discover the amount of error, if any, in its alignment. To do this, turn and thread on centers a plug that will fit the tapped hole firmly, without bottoming in the hole. The diameter of the threaded portion of this plug should be less than the diameter of the tap which was used to finish the thread in the arbor. Use another lathe in making this plug, so as to avoid disturbing the setting of the arbor. Screw this plug into the hole, then revolve the arbor and plug slowly, the arbor still being in the steadyrest. Place an indicator against the end of the plug at K in order to determine how much the end is out of true. The eccentricity of the center L of the plug is half the oscillation of the indicator pointer.

Referring to the diagram, in Fig. 4, showing center-lines only, the distance LB_1 is the eccentricity of the plug at L. The line ABB_1 is the center-line of the arbor, and A_1BL is the center-line of the tapped hole. It is evident that if the center A coincided with A_1 the arbor would be perfect. But we shall assume that the indicator shows an eccentricity of 0.002 inch when placed at K. This means that the oscillation of the pointer is 0.004 inch. Suppose that the center-lines intersect at a point B. This point can be determined approximately by proportion, after making one more test with the indicator. Move the indicator to point M, and the oscillation at this point will be more or less than at K. The point B is located where the oscillation would cease. In the diagram it is shown near the end of the arbor, though it might be far outside. In this case, A_1 would be the correct position for the center A. The distance AA_1 is the eccentricity between the center A and its correct position A_1. This distance may be determined by proportion.

$$AA_1 \div LB_1 = AB \div BB_1$$
$$AA_1 = \frac{LB_1 \times AB}{BB_1}.$$

Suppose $AB = 6$ inches and $BB_1 = 2$ inches, and that, as we assumed before, $LB_1 = $ 0.002 inch. Then

$$AA_1 = \frac{0.002 \times 6}{2} = 0.006 \text{ inch.}$$

Hence, if center A be drawn toward A_1 0.006 inch, it will be in its proper place, *i.e.*, at A_1.

After correcting the arbor, as suggested above, it should again be tested. This time, however, remove the steadyrest and run the tailstock center up to the end of the plug at L. Rotate the arbor and plug on the centers A_1 and L, and test by placing the indicator at M. If we still find that the indicator shows an error, make the necessary correction by slightly scraping the center A_1. When correct, no error should be shown by the indicator.

Now that we are assured that the centers are right, we can proceed to finish the arbor. If the plug is stiff enough, finish turn the arbor on the centers A_1 and L. If the plug is frail, simply use it to take a light cut on the end E. Then place the steadyrest on this new surface of E, and remove the plug. Bore carefully the taper shown at T. Remove the steadyrest and mount the arbor on its own centers, one center being formed by the taper T. Then, finally, finish the arbor on the outside.

A slight variation from the above method is advisable when the threaded hole is large in proportion to the rest of the arbor. In this case the plug should be made in a chuck, and the arbor screwed onto it, and the center A_1 determined by spotting.

Seasoning Hardened Steel. — It is a well-known fact that hardened pieces of steel will undergo minute but measurable changes in form during a long period of time after the hardening has taken place. These changes are due to the internal stresses produced by the hardening process, which are slowly and gradually relieved. In order to eliminate slight inaccuracies which might result from these changes, steel used for gages and other tools requiring a high degree of accuracy, is allowed to season before it is finally ground and lapped to the finished dimensions. The time allowed for this seasoning varies considerably among different toolmakers and also depends upon the form of the work and the degree of accuracy which is necessary in the finished product. Some toolmakers rough grind the hardened part quite close to the finished size and then allow it to season or "age" for three or four months, and in some cases, a year or more. According to one manufacturer, it requires a period of about eight years for a piece of steel to thoroughly season unaided. Experiments made in Germany indicate that changes in the dimensions of hardened steel take place sometimes during a period as long as three years. In order to find a method of overcoming the delay due to seasoning hardened steel, experiments showed that by heating the hardened steel in oil at a temperature of about 300° F. for a period of ten hours, the internal stresses which caused the changes were practically eliminated.

Another method which has been applied to gage making in order to hasten the seasoning period is to draw the hardened gage to a straw color after it has been roughed out. This method may be objectionable in that it leaves the gage too soft. Still another method of seasoning steel is to remove all scale and place the steel part on the base of an electrical generator where it is subjected to constant vibration and ever changing temperature which reduces the seasoning period to about six weeks. What is known as the hot- and cold-water method is also commonly employed for seasoning or aging steel. The steel so treated is first roughed out quite close to the finished size and is then immersed in a bath of boiling water which causes it to expand; then it is dipped into ice water, and this operation is repeated perhaps a hundred times. The repeated expansions and contractions cause the molecules to settle into permanent form. Steel that has been carefully annealed, changes very little if left unhardened. There is also very little change in parts which are only partially hardened, especially if the hardened parts are small in proportion to the size of the work.

Seasoning Cast Iron. — Castings will often change their shape slightly after being planed, especially if the planed surface represents a large proportion of the total surface. To prevent errors from such changes, castings are sometimes allowed to season for several weeks or months after taking the roughing cuts and before finishing. The change in shape is due to a re-adjustment of internal stresses thrown out of balance by the removal of the surfaces. These forces set up by irregular cooling of the metals in the foundry, are gradually neutralized during the seasoning period. They may be partially eliminated by re-heating, tumbling, and rapping; repeated dippings in hot water also has an accelerating tendency.

A common method of avoiding the long seasoning period is to anneal the castings. This practice is followed by at least one manufacturer for relieving the stresses in the beds of measuring machines. The Jones process furnace is used and the work to be annealed is kept in a closed receptacle. This prevents oxidization and the work comes out as bright as it goes into the furnace. The measuring machine beds are of box section and about three feet long. One-sixteenth inch is allowed for the finishing cut after annealing. The planing is practically all on one side and none of the castings have warped in the annealing to such an extent that they would not finish up properly in the subsequent re-planing. The special skill in connection with annealing lies in choosing the proper points for supporting the castings and heating it to a low red heat only; it has not been found necessary to use a higher heat in order to eliminate all the internal strains. At a higher heat, the casting would undoubtedly sag out of shape by its own weight. Experiments have shown that the release of cooling strains in iron castings may be very greatly accelerated without annealing, by subjecting the castings to repeated shocks or vibrations while cold. The effect of this treatment seems to be similar to that obtained by heating the castings.

Cutting Teeth in End and Side Mills. — When the end teeth of an end mill or the side teeth of a side mill are being cut, it is

necessary to set the dividing head at an angle as shown at *A*
and *B*, in Fig. 5, in order to mill the lands or tops of the teeth
to a uniform width.   To determine this angle α, multiply the
tangent of the angle between adjacent teeth (equals 360 ÷
number of teeth required) by the cotangent of the cutter angle;
the result is the cosine of angle α at which the dividing head
must be set.

Example. — An end mill is to have 10 teeth.   At what angle
should the dividing head be set, assuming that a 70-degree
fluting cutter is to be used?

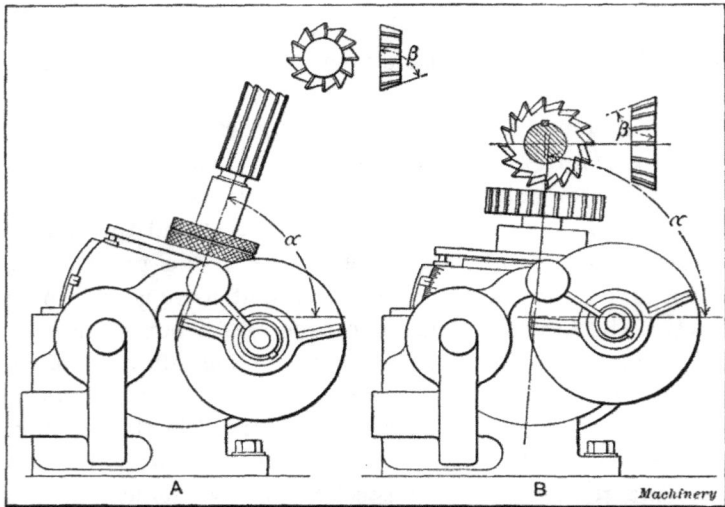

Fig. 5.   (A) End Mill set for Milling End Teeth.   (B) Milling Side
Teeth of Side Mill

The angle between adjacent teeth equals 360 ÷ 10 = 36 degrees.   The tangent of 36 degrees is 0.7265 and the cotangent
of 70 degrees is 0.3639; therefore, cosine of angle α to which
the dividing head should be set equals 0.7265 × 0.3639 = 0.2643.
The angle whose cosine is 0.2643 is 74 degrees 40 minutes;
hence the dividing head would be set to this angle as indicated
at *A*, Fig. 5.

The angle of elevation for cutting the side teeth of a side
mill would be determined in the same way.   Sketch *B* shows a

dividing head set for milling the side teeth of a side mill using a 70-degree cutter, the angle α being approximately $85\frac{1}{2}$ degrees.

Fluting Angular Cutters. — When milling the teeth of angular cutters or fluting taper reamers with a single-angle cutter, the angle at which to set the dividing head above the horizontal position equals the angle $\gamma - \delta$ (see Fig. 6); the tangent of $\gamma = \dfrac{\cos \theta}{\tan \beta}$, and the sine of $\delta = \tan \theta \cot \phi \sin \gamma$. These formulas can be expressed as a rule, although the latter is rather cumber-

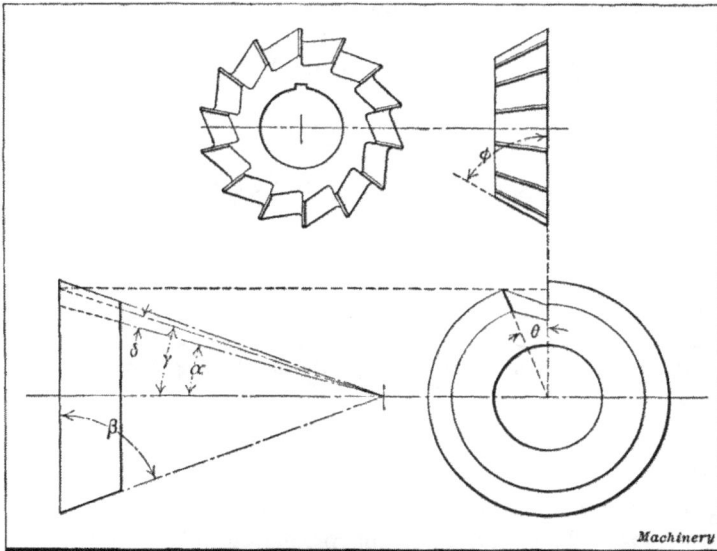

Fig. 6. Angles Involved in Calculation for Determining Position of Index Head when Milling Teeth in Angular Cutter

some. To determine the angular setting of the indexing head, divide the cosine of the angle θ (Fig. 6) between adjacent teeth (equals 360 ÷ number of teeth required) by the tangent of the blank angle β; the quotient is the tangent of angle γ. Next multiply the tangent of angle θ between the teeth, by the cotangent of the grooving cutter angle ϕ, and then multiply the result by the sine of angle γ; the product will be the sine of angle δ. Subtract angle δ from angle γ to obtain the angle α at which the dividing head should be set.

Cutter Grinding. — The best way to hold cylindrical milling cutters for sharpening is to mount them on a sleeve made for this purpose. These sleeves, which are illustrated in Fig. 7, are used in many machine shops. They slide on a round bar held in the cutter grinder, and if the bar is straight, it is evident that the teeth of the cutter will be ground parallel. Cutters can, of course, be placed on an ordinary mandrel or arbor held between the centers of the cutter grinder. If this method is used it is well to be certain that the mandrel runs true and that the centers are in line with the ways of the machine, as otherwise the teeth of the cutter will not be parallel. Since it takes a great deal of cutting and trying to determine this, it is best

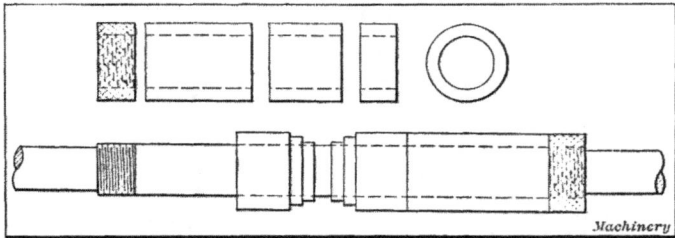

Fig. 7. Arbor and Sleeves used for Holding Milling Cutters when Grinding Teeth

to use the other method when possible, as it is more rapid and reliable.

Care should be taken in keeping cutters round while sharpening them. The first tooth sharpened should be marked with chalk before starting, and each tooth should be moved past the wheel with a steady, but not necessarily slow, motion. Care should be taken not to remove too much stock at one cut; if crowded, the tendency will be to burn the work and cause the wheel to wear away too rapidly. The rapid wearing of the wheel will leave the cutter considerably out of round. Under favorable conditions a cut of 0.003 inch can be taken without burning the teeth or wearing the wheel away enough to cause the cutter to be out of round more than 0.001 inch. A slight error of this kind is corrected in the finishing cut. The operator should finish up with a light cut, say 0.001 inch deep, and before removing the cutter, he should make sure that it is round. This is easily determined by the manner in which the teeth "spark" on the wheel.

Wheels for Sharpening Milling Cutters. — Grinding wheels for cutter sharpening should be of a medium-soft grade and not too fine — never finer than 60 grit. Fine wheels cut slowly and tend to burn the teeth. Wheels for sharpening cutters made of high-speed steels can be a little coarser than those used on carbon steel. If the wheel is too soft, it will wear rapidly, which makes it difficult to keep the cutter round while sharpening it. This difficulty can be overcome by using a wheel at least $\frac{3}{4}$-inch wide, instead of the $\frac{3}{8}$- or $\frac{1}{2}$-inch wheels commonly used. A wide soft wheel will last as long as a narrow one of harder grade, and being softer, it tends to eliminate the danger of burning. For sharpening ordinary milling cutters, a wide wheel will not be especially inconvenient, as generally there is plenty of room. The following "Aloxite" wheels have given good results for cutter sharpening operations: For carbon steel mills and cutters, 50-O; for high-speed steel mills and cutters, 40-O; for carbon-steel formed cutters, 50-P; for high-speed steel formed cutters, 40-P.

To obtain the best results, these wheels should run at a surface speed of about 5000 feet per minute. Ordinary milling cutters may be sharpened either by using the periphery of a disk wheel or the face of a cup wheel. The latter has the advantage of grinding flat lands, whereas the periphery of a disk wheel leaves the teeth slightly concave, thereby causing them to become dull sooner than they would if the lands were flat. However if the disk wheel is not under 3 inches in diameter, the teeth will not be concaved enough to cause trouble; moreover the disk wheel will grind faster.

Rotation of Wheel Relative to Cutter. — When grinding the teeth of cutters, reamers, etc., in a regular cutter grinding machine, the wheel can be revolved either against the cutting edge of the tooth as shown by arrow B (Fig. 8) or away from it, as indicated by arrow C. The cutter can be presented in either way, by swiveling the work around the stationary column of the machine, to either side of the wheel. By revolving the wheel against the tooth, as at *B*, a keen edge will be obtained without forming a burr, and there is also less danger of drawing the temper, thus enabling the grinding to be done more rapidly. Care must be taken, however, to hold the work securely against the tooth-rest as otherwise the wheel may draw the cutter away from the rest and score the tooth. There is also danger of the wheel being broken. Rotating the wheel as shown by arrow *C* is the safer method, as the wheel then holds the tooth against the rest. When the wheel is rotated in this direction, however,

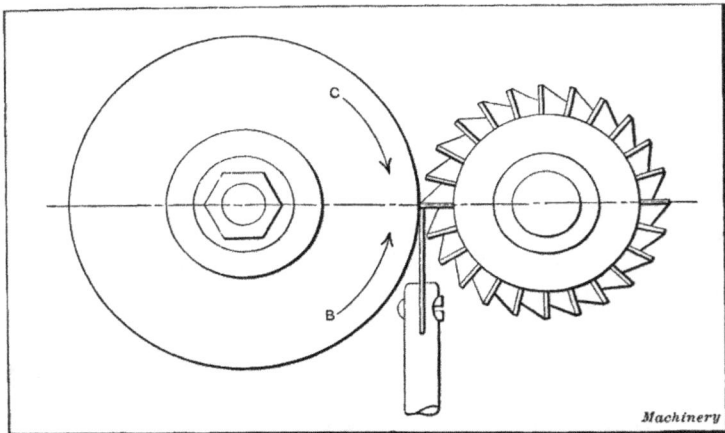

Fig. 8.   Rotation of Grinding Wheel Relative to Cutter may be as shown by Arrow B, although Opposite Direction is usually considered Safer

a slight burr is left on the cutting edge, which should be removed with an oilstone.

Location of Tooth-rest for Cutter Grinding. — The tooth-rest which is used to support the cutter while grinding the teeth, should be located with reference to the nature of the work. When grinding a cylindrical cutter having helical teeth, the tooth-rest must remain in a fixed position relative to the grinding wheel. The tooth being ground will then slide over the tooth-rest, thus causing the cutter to turn as it moves longitudinally, so that the edge of the helical tooth is ground to a uniform distance from the center, throughout its length. Obviously, if the tooth-rest were attached to the machine table and was fixed relative to the cutter, it would be impossible for the wheel to follow the helical teeth. When grinding a straight-fluted cutter, it is also preferable to have the tooth-rest in a fixed position relative to the wheel, unless the cutter is quite narrow, because any warping of the cutter in hardening will result in inaccurate grinding if the tooth-rest moves with the work. The tooth-rest should be placed as close to the cutting edge of the cutter as is practicable, and bear against the face of the tooth being ground. When the tooth-rest is fixed relative to the wheel, it should be somewhat wider than the wheel face so that the cutter will have a support before it reaches the

Fig. 9. Correct and Incorrect Positions of Tooth-rest when Sharpening Angular Cutters

wheel and also after it has been traversed past the wheel face. The end of the tooth-rest should have an even bearing upon the tooth being ground. Narrow tooth-rests may be used when they are attached to the table, and remain fixed relative to the work.

Sharpening Angular Cutters. — In sharpening cutters of this type, the tooth-rest should be set exactly on the center as indicated at *A*, Fig. 9, so that the cutter will be ground to the angle shown by the graduations on the machine. The angle for clearance, in this case, is obtained either by raising the wheel or lowering the swiveling head, depending upon the design of the machine. The tooth-rest should never be set below the center, as at *B*. Many machinists and toolmakers overlook

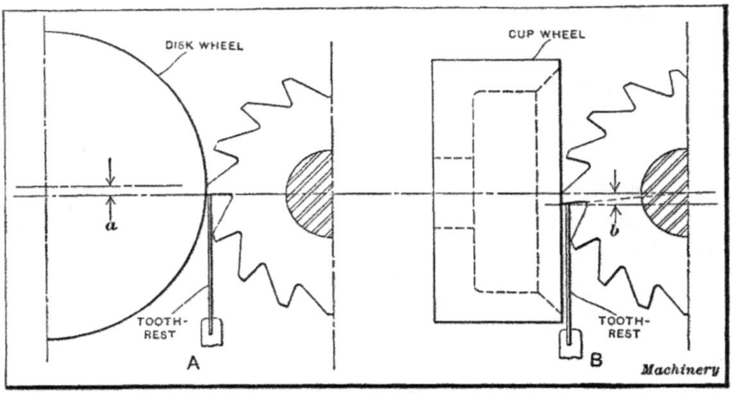

Offset of Disk Wheel for Grinding Clearance

Wheel diameter....	3½	3¾	4	4¼	4½	4¾	5	5¼	5½	6
Offset *a*, 5° clearance....	5/32	5/32	11/64	3/16	13/64	13/64	7/32	15/64	15/64	17/64
Offset *a*, 7° clearance....	7/32	15/64	1/4	17/64	9/32	19/64	5/16	21/64	11/32	3/8

Offset of Cup Wheel for Grinding Clearance

Cutter diameter...	7/8	1	1⅛	1¼	1½	1¾	2	2¼	2½	3
Offset *b*, 5° clearance....	0.037	0.044	0.050	0.055	0.066	0.077	0.088	0.099	0.110	0.132
Offset *b*, 7° clearance....	0.052	0.060	0.067	0.075	0.090	0.105	0.120	0.135	0.150	0.180
Cutter diameter...	3½	3¾	4	4½	5	5½	6	6½	7	8
Offset *b*, 5° clearance....	0.154	0.165	0.176	0.198	0.220	0.242	0.264	0.286	0.308	0.352
Offset *b*, 7° clearance....	0.210	0.225	0.240	0.270	0.300	0.330	0.360	0.390	0.420	0.480

Fig. 10. Methods of obtaining Clearance on Cutter Teeth when using Disk and Cup Wheels — Tables giving Amount of Offset for Different Clearance Angles

this point, which is often the cause of angular cutters being inaccurate and unfit for close work.

Clearance Angle for Teeth. — Milling cutters usually have from 5 to 7 degrees of clearance. The practice in some shops, where special roughing and finishing cutters are used, is to give the roughing cutters a clearance of 7 degrees and finishing cutters, 5 degrees. Excessive clearance causes chattering when milling, and the teeth become dull quickly. The Brown & Sharpe Mfg. Co. recommends a clearance of 4 degrees for plain milling cutters over 3 inches in diameter, and 6 degrees for those under 3 inches. The clearance of the end teeth of end mills should be about 2 degrees, and it is advisable to grind the teeth about 0.001 or 0.002 inch low in the center so that the inner ends of the teeth will not drag on the work. The clearance angle is regulated, when grinding, by setting the center of the grinding wheel slightly above the center of the cutter, or by adjusting the tooth-rest slightly below the center. When grinding with a plain disk wheel, the clearance may be obtained by setting the center of the grinding wheel above the center of the work as shown at A, Fig. 10. When using a cup wheel, the clearance may be obtained as indicated at B. The grinding wheel and work centers are at the same height and the tooth-rest is set below the centers as shown. The accompanying tables give the amount of offset a and b (see Fig. 10) for disk and cup wheels, respectively.

Sharpening Formed Cutters. — Formed cutters should be ground radially or so that the faces of the teeth lie in planes passing through the axis of the cutter. The teeth should also have the same height to insure each tooth doing an equal amount of work. When setting up the grinder for formed cutters, the grinding side of the wheel should run true and be in line with the centers. If the wheel is set off center and the teeth are not ground radially, the cutter will not mill the desired shape. One way to sharpen form cutters is on a surface grinding machine, using for this purpose the dividing centers with which it is generally equipped. Another, and perhaps a more convenient, way is with a form cutter grinding attachment as shown in Fig. 11. The accurate spacing of the teeth must be maintained under all conditions, or else all of the teeth will not cut; and as form cutters have comparatively few teeth, it is best to have all of them in working condition. This makes it necessary to remove the same amount from each tooth (providing the spacing is accurate) every time the cutter is ground, regardless of whether all of the teeth are dull or not. If some of the teeth are dull and some sharp it may be caused by the teeth having been inaccurately spaced in a previous grinding operation, or

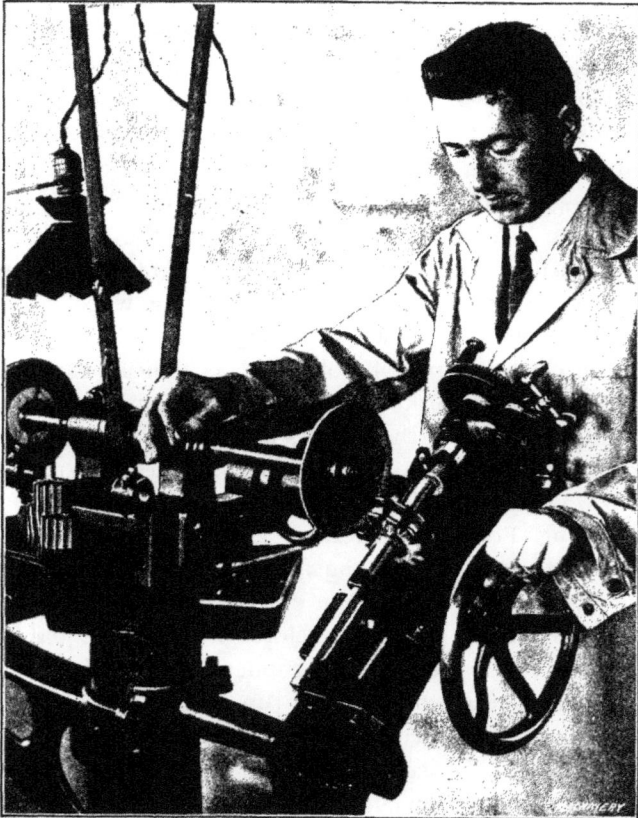

Fig. 11.   Grinding Form Cutters by Means of a Form Cutter
Grinding Attachment

it may be that the cutter has been used on an arbor that did not run true.

After the wheel has been set, its position relative to the centers should not be disturbed. For this reason the cutter should be brought to the wheel by slightly adjusting the arm that carries

the index pin.   If the wheel has to be dressed during the grinding operation it should be set central again before proceeding
with the work.   Each tooth should be ground evenly and carefully with a comparatively light cut, as a heavy cut will burn
the teeth.   The operator should bear in mind that it takes time
and patience to properly sharpen form cutters.   Never under
any circumstances should the work be done by hand.

The most important of the formed milling cutters are those
used for cutting gears, and it is very essential that the
original shape be maintained throughout their working life,
because slight deviations from the original shape that would
pass unnoticed with other formed cutters become very apparent
in cut gears.   Inaccurate tooth shapes and lack of concentricity
of the pitch circle with the mounting shaft, are the two chief
causes of unsatisfactory action of gears cut with the common
gear cutter.   To grind a gear cutter so as to preserve the original
shape is not an easy matter unless better means are provided
than are found in most shops.   In the first place the cutting
face must be ground radial; any inclination of the face ahead
or back of the radial line changes the projected shape, and as
it is that which defines the outline of the cut, the result, of
course, is a departure from the established tooth shape.   To
grind radial faces with a cutter grinder is comparatively easy,
and if this were the only requirement to produce perfect working cutters the problem were easily solved, but the heights of
the teeth must be equal.   This, too, could be readily accomplished by using an indexing fixture if the cutters were truly
round and evenly spaced.   These conditions are seldom real-
ized, however.   Changes of shape in hardening throw the teeth
out of even spacing, and if ground on an accurate indexing
fixture some of the teeth will be so short that only a few do the
cutting.   The result is rough work and heating of the cutter
which reduces the quality and quantity of gears cut.

A cutter grinder recently put on the market employs a principle of indexing which is unknown to most mechanics.   The
teeth are indexed by a finger resting on the *back* of the formed
part, an indicator being provided which enables the operator

to gage the position of each tooth with reference to its own shape as well as to the axis of the cutter, within a thousandth inch or less; therefore each tooth face is ground truly radial and all the teeth are ground to the same height.

Sharpening End Mills. — When re-sharpening, end mills should never be held between centers, even if it appears to be the easier way. They should always be located from the shank,

Fig. 12.   Sharpening Side Milling Cutters

because end mills are almost always out of true after having been used, and for this reason the teeth cannot be brought concentric with the shank if the mill is held between centers. All makers of cutter grinders should provide suitable collets for this purpose to fit the swivel head of the machine. The swivel head must be in line with the ways of the machine, otherwise the teeth will not be parallel with each other. The work should be calipered at both ends with a micrometer to make sure that it is straight. The same wheels can be used for this work that are used for sharpening ordinary milling cutters, and the process is much the same. In sharpening spiral mills or cutters, the operator should see that the guiding finger bears evenly on the tooth as otherwise it will be difficult to keep the teeth parallel.

Sharpening the Side Teeth of Cutters. — The side teeth of milling cutters require very little attention after they have been backed off when the cutter is new. They are generally sharpened by being held on a straight arbor with one end turned down so as to bring the cutter up to a shoulder. A screw and washer hold the cutter in position. When arranged as shown in Fig. 12, the arbor for holding the cutter rests in a V in the cutter

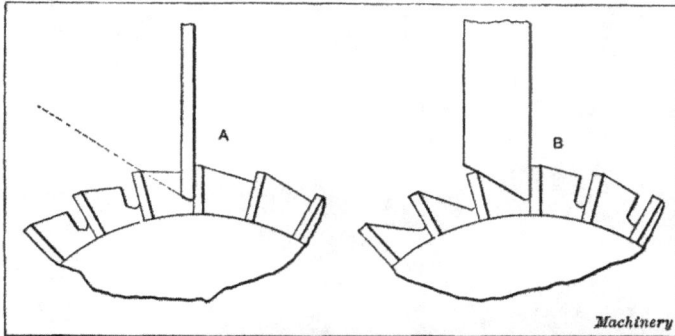

Fig. 13. The Two Steps taken in Re-cutting Milling Cutters by Grinding

sharpening head which is set over to an angle of five degrees to give the proper clearance to the teeth. The teeth are sharpened by feeding them past the face of a disk or cup wheel. A wheel for this purpose should be very free cutting; otherwise the teeth will be burned and thus ruined. As a general rule the side of a wheel is never as free cutting as the periphery, and for this reason the depth of cut should seldom be over 0.002 inch deep. It is preferable to run the wheel towards the edge of the tooth as this helps to overcome the danger of burning.

Re-fluting Worn Cutters by Grinding. — When the lands on the teeth of a milling cutter become too wide and the flutes which form the teeth become too small, as the result of repeated grinding on the tops of the teeth, the cutter can be restored

without annealing it, by grinding the backs of the teeth. All
that is required is an ordinary cutter grinding machine, such as
is found in nearly every tool-room, and wheels of the proper
materials and grades. When re-cutting worn-out milling cutters, about the same methods are used as when cutting new

Fig. 14.   Re-cutting Milling Cutters in Brown & Sharpe No. 3 Cutter
Grinder

cutters, except that grinding machinery and abrasive wheels
are used in place of the milling machine and milling cutters.

The method followed can be readily understood from Fig. 13.
This method applies to both peripheral and side teeth. The
teeth are first gashed out as shown at A, with a thin vulcanite
wheel, the object of this gashing being to preserve the corner

of the wheel used in the re-cutting operation. At *B* is shown the re-cutting operation, the face of the wheel being trued to the proper angle. To re-cut milling cutters in one operation is not practical, as a hard and, therefore, slow-cutting wheel would have to be used to preserve the corner of the wheel. With a thin elastic wheel of just the proper grain and texture for the gashing operation, and a fast-cutting vitrified wheel to cut away the superfluous stock between the bottoms of the gashes and the points of the teeth, very economical and satisfactory results can be obtained with a little practice.

Fig. 14 shows how the preliminary gashing operation illustrated at *A*, Fig. 13, was done in a B. & S. cutter grinder. The cutter was held on an arbor mounted between the centers, and the teeth were gashed out one after another. For this purpose, an aloxite vulcanite wheel 7 inches in diameter, $\frac{1}{8}$-inch face, $1\frac{1}{4}$-inch hole, 40 grit, V-K-9 bond was used. This wheel was run at a speed of 4370 R.P.M. A higher speed would be practicable, as a vulcanite wheel must be run at a very high speed to give high efficiency. Contrary to the general rule, wheels with vulcanite bond do not burn the work when run at high speeds. With the cutters in question, the gashing was made $\frac{1}{8}$ inch deep. One tooth was done at a time, with three cuts, one after another, so that each cut was approximately 0.042 inch deep. With this cut no evidence of burning was noticeable. The direction of the wheel rotation and feed should be the same. When fed in the opposite direction, which on first thought appeared to be the more practical way, it was found that the wheel had a tendency to burn, in breaking through. The object of this gashing was to preserve the corner of the wheel used for the second operation, which consisted of grinding away the superfluous stock between the bottom of the gashes and the points of the teeth.

For this final operation, the cutter was held as before and the teeth were ground out one at a time by being fed under a wheel having its face trued to the proper angle. An aloxite wheel 5 inches in diameter, $\frac{3}{8}$-inch face, $1\frac{1}{4}$-inch hole, 50 grit, O grade, D496 bond was used, the speed being 3500 R.P.M. The wheel was set central, and in re-cutting, the face of each tooth was brought to within a few thousandths inch of the back or flat side of the wheel. The cutter was then fed back and forth under the wheel taking a chip of approximately 0.002 inch at a time, until the land of the tooth was of the proper dimension, or about $\frac{1}{32}$ inch wide. The cutter was then turned just enough to bring what was left of the original face of the tooth to bear on the back side of the wheel. This removed the small amount of stock not taken out by the first operation, and matched up the side and face teeth.

Re-fluting Slotting Saws by Grinding. — Figs. 15 and 16 illustrate how the teeth of six worn slotting saws were re-cut to shape

Fig. 15.   The Gashing Operation

by grinding. The worn saws were first placed on an ordinary work arbor such as used in turning up thin collars, etc., paper washers being placed between the cutters to compensate for the side taper. They were then trued up in a cylindrical grinder, the old teeth being partially ground away. This work was done dry with an aloxite wheel 6 inches in diameter, $\frac{1}{2}$-inch face, $1\frac{1}{4}$-inch hole, 50 grit, M grade, D496 bond. The work was then placed between the centers of the form cutter attachment of a Brown & Sharpe No. 3 cutter grinder, and the teeth gashed out to the required depth. This operation is shown in Fig. 15. The wheel used for this purpose was aloxite (vulcanite), 7 inches in diameter, $\frac{1}{8}$-inch face, $1\frac{1}{4}$-inch hole, 30 grit, V-K-9 bond, run at a speed of about 4200 R.P.M. In this operation the wheel loss was only 0.008 inch. The depth of cut was 0.003 inch for each stroke, the graduations on the cross-feed screw being relied on for the correct depth. The work was fed under the wheel with a fairly rapid motion, one tooth being gashed to its full depth at a time.

The next operation, which consists of cutting the new teeth, is shown in Fig. 16. This operation is practically the same as the gashing operation, the difference being that a wide wheel is used,

Fig. 16. Grinding the New Teeth

its face being trued to the correct angle, as shown at B in Fig. 13. The wheel used for this purpose is aloxite, 6 inches in diameter, $\frac{3}{4}$-inch face, $1\frac{1}{4}$-inch hole, 40 grit, O grade, D497 bond, run at a speed of about 4200 R.P.M. The teeth were fed one at a time under the wheel, taking a cut of 0.002 inch at each stroke until the land of the tooth was of the correct thickness, or about $\frac{1}{32}$ inch. The wheel loss in this operation was $\frac{1}{64}$ inch. The teeth were then backed off as shown in Fig. 17. For this purpose an aloxite wheel 5 inches in diameter, $\frac{3}{8}$-inch face, $1\frac{1}{4}$-inch hole, 50 grit, O grade, D496 bond, was used, the speed being 3650 R.P.M.

Re-fluting Shell End Mills by Grinding. — In Figs. 18 and 19, the operation of re-fluting the end teeth of shell end mills is illustrated.   Before re-cutting, the operator should first decide on the correct angle for the teeth, and then true the face of both wheels to this angle.   In the gashing operation the cut should be carried to a depth where a line drawn parallel to the face of the wheel

Fig. 17.   Grinding the Clearance

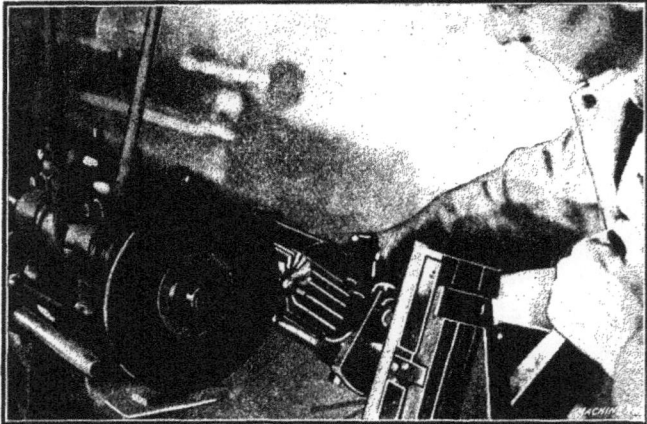

Fig. 18.   Gashing End-mill Teeth

comes nearly to the edge of the tooth.   (See *A* in Fig. 13.)   In re-cutting end teeth it is necessary to use a machine equipped with a head that can be set at any angle in order to bring the lands of the teeth parallel.   Figs. 18 and 19 show a Brown &

Sharpe No. 3 cutter grinder, equipped with an end-mill grinding attachment. In setting the head to the correct angle the cut-and-try method can be used, or the angle can be figured out when

Fig. 19. Grinding End-mill Teeth to Shape

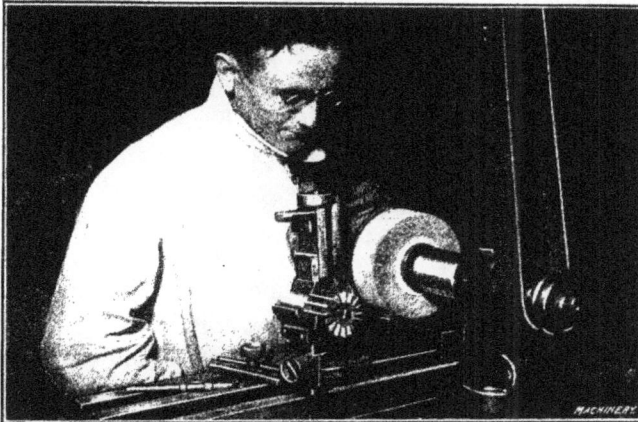

Fig. 20. Grinding the Clearance on End-mill Teeth

the operator has decided on the angle for the wheel face and knows the number of teeth to be cut.

To determine the angle at which the cutter arbor should be inclined from the vertical, multiply the tangent of the angle between adjacent teeth (this angle equals 360 ÷ number of teeth) by the cotangent of the wheel face angle; the result is the cosine of the angle at which the cutter arbor should be set.

The gashing operation is shown in Fig. 18, the teeth being fed one at a time under the wheel with a cut of 0.002 inch, until the correct depth is reached. This is easily determined for all of the teeth by the graduations on the cross-slide screw, after the proper depth for the first tooth is decided on. In re-cutting four mills having sixteen teeth each, the loss of the gashing wheel was only $\frac{1}{64}$ inch. The re-cutting operation is shown in Fig. 19, the teeth being ground until the lands were of the correct dimension, or $\frac{1}{32}$ inch. The wheel loss in this operation was $\frac{1}{8}$ inch for re-

Fig. 21. A Typical Formed Cutter

cutting four mills having sixteen teeth each. As the width of the land is relied on in this case to determine the proper depth of the cut, the wheel loss is of no consideration, provided the wheel wears true, thus preserving approximately the correct angle. In this operation the most satisfactory results were attained with a soft coarse wheel, run at a comparatively high speed. When finer wheels in harder grades were used the tendency was to cut slow and burn.

After re-cutting the teeth were backed off as shown in Fig. 20, this work being done on a Walker grinder equipped with a universal cutter grinding head. The wheel used for this operation was aloxite (cup), $5\frac{1}{2}$ inches in diameter, 2 inches face, $1\frac{1}{4}$-inch hole, $\frac{3}{4}$-inch back, $\frac{3}{8}$-inch wall, 50 grain, O grade.

Making Formed Cutters. — Milling cutters of irregular shape that can be sharpened without altering their contour (an example of which is shown in Fig. 21), are generally called formed cutters. They are used for milling irregular surfaces on gun and sewing machine parts and on other work of like nature. Perhaps the first formed cutters to be commonly used were the gear-cutters known to every mechanic. Satisfactory formed cutters can be procured from any cutter manufacturer at a moderate cost, but as several weeks often elapse before special cutters are delivered, many shops are equipped to make their own cutters.

In explaining the various steps in the making of a formed cutter we can use the tool illustrated in Fig. 21 as an example. As this is a male cutter it is formed and backed off with the female former shown at the left in Fig. 22. As this former has to have a clearance of at least 15 degrees, with the original lay-out accurately preserved, it is necessary to form it at an angle by means of the master shown at the center in Fig. 22. As this

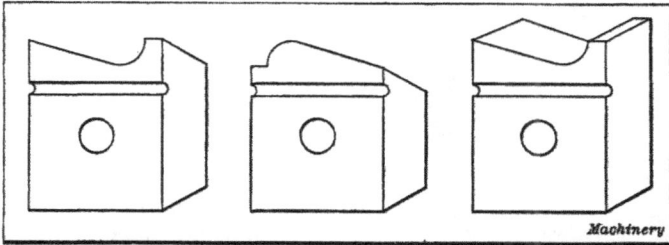

Fig. 22. Former and Master Tools for Producing Formed Cutters

master has a convex surface that would be difficult to work out accurately by hand, a second or female master is made. This is shown at the right, and it is an exact counterpart of the shape we wish to mill, worked out in a block of steel without any clearance. This piece is hardened and drawn to a medium straw color.

To make the masters and former without distorting the contour, and at the same time give the former the necessary clearance, the angle block and tool shown in Fig. 23 are used. The angle block is strapped to the platen of the shaper and the female master is fastened to the angle tool. First, however, the outline is worked to shape as near as possible with regular tools, the female master being used only for finishing. As the illustration shows, it works with a drag cut which necessarily means slow cutting. It will also be seen that the faces of the two masters are parallel; this is why the proper contour can be maintained, notwithstanding the fact that clearance is imparted to the male master. This master is now hardened and drawn and the same process repeated in making the female former. The masters and former are sharpened by grinding their tops or faces. The slight groove shown is for the corner of the wheel to run into. They should be sharpened on the surface grinder, as their faces must remain flat in order to retain the correct shape at the cutting edge. In finishing the master and former on the angle block, it is necessary to use a very slow cutting speed to avoid chattering. As a very smooth surface is desired, good lard oil should be used as a cutting medium.

Formed cutters can be made of bar stock chucked out and turned up in the usual way. About 0.003 inch should be left in the hole for finishing by grinding after hardening, and it is a good plan to relieve the sides as shown in Fig. 21, as this eliminates a certain amount of unnecessary grinding. The hole should be chambered and

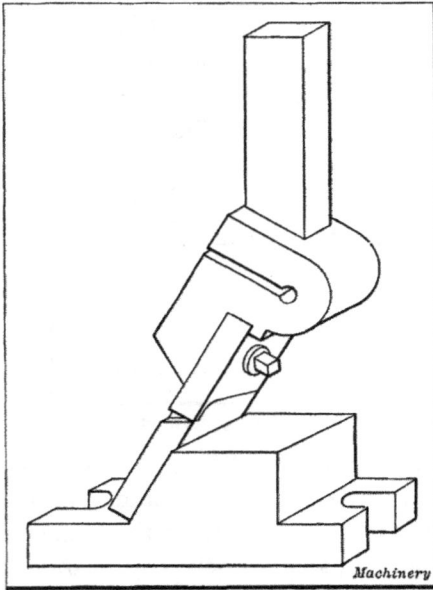

Fig. 23.   Angle Block and Tool used for
Making the Former and Master Tool

provided with a keyway. The cutter is turned to the required shape with the forming tool previously made. This is done in the lathe, the forming tool being fastened to the tool-holder shown in Fig. 24. As will be seen, this tool-holder and also the one shown in Fig. 23 are of the "goose-neck" type, to obviate chattering. The cutter is now ready for fluting or gashing. In this case, a 30-degree angular cutter rounded at the point may be used. After this operation is finished, the cutter is ready for relieving, or, as it is more commonly called, backing off.

Relieving the Formed Teeth. — There are a number of different attachments for backing off the teeth of formed cutters which may be applied to the lathe and milling machine. The device shown in Fig. 25 is simple to construct, easy to set up and it gives very good results. Briefly described, it consists of a slid-

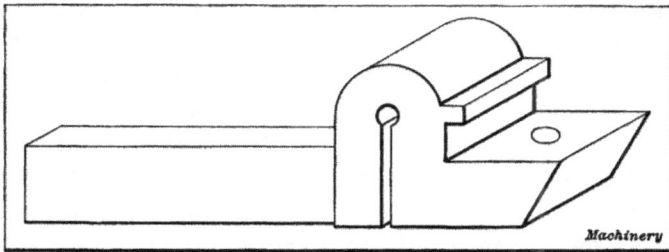

Fig. 24. Gooseneck Tool-holder on which Former is Mounted for Machining Formed Cutters

ing base mounted in a bed that is strapped to the milling machine platen. A strong spring thrusts the base toward the spindle and cam. The cam is mounted on a cast-iron holder that screws onto the nose of the machine spindle, while the cutter is mounted on a stub arbor. The forming tool, shown at the left in Fig. 22, is

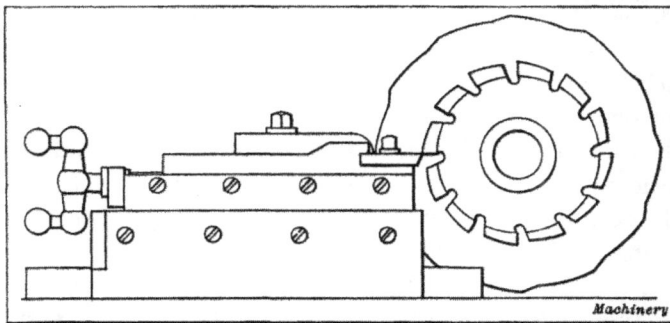

Fig. 25. Milling Machine Relieving Attachment for backing off Formed Cutter Teeth

fastened to a sliding head that is actuated by a screw which regulates the depth of cut in backing off the cutter. The details of the relieving attachment are shown in Fig. 26, which illustrates an end elevation and longitudinal section. As each cam has a fixed number of teeth, three are generally provided for relieving cutters having ten, twelve and fourteen teeth, respectively. It is very necessary that these cams be accurately made as otherwise faulty cutters will result. The method employed in laying out and machining them will be explained later.

In setting up for the backing-off operation, it is necessary that the cutter be set so that the teeth are in the correct relation with the corresponding cam rises. To secure this adjustment, an arbor without a tang is used. A line is scribed on the cam meeting the high points of two rises that are opposite. The arbor is placed loosely in position, the cam set central by means of the scribed lines and a surface gage, and the arbor adjusted so that the cutter tooth does not meet the former within $\frac{1}{16}$ inch. The arbor is now driven solidly into position. The object of the

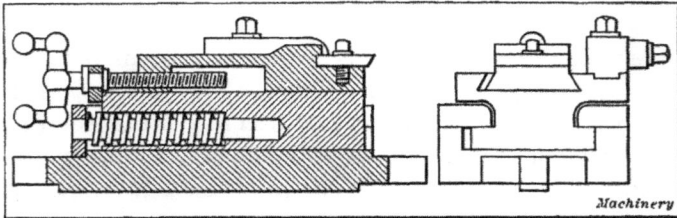

Fig. 26. End Elevation and Sectional View of Relieving Attachment shown in Fig. 25

$\frac{1}{16}$-inch leeway is to make sure that the former is moving toward the cutter before the tooth is reached.

When backing off the teeth, a very slow speed should be used and the work should be lubricated with lard oil. Light cuts should be taken, as heavy cuts are likely to leave scores which are reproduced by the finished cutter. After the teeth have been backed off, the former should be reground and the top whetted on an India oilstone wet with gasoline. It should be examined under a glass to make sure that it has a clean sharp edge without scores. It is then placed in position again and a light finishing cut taken. It is necessary that the top of the former be set central with the cutter, for if it is above or below, an incorrect contour will be formed.

The cutter is now hardened and drawn in the usual way. After hardening, the sides should first be ground. A Heald piston-ring grinder is an excellent machine to utilize for this purpose, but if one is not available, the universal grinder can be used. After grinding the sides, the cutter is strapped to the faceplate of the grinding machine, the hole trued up with an indicator and ground to the desired size. The next operation is to grind the faces of the teeth. While this is a simple operation, care should be exercised to see that the teeth are equally spaced, for while they probably were equally spaced after cutting, the hardening process may have distorted them slightly.

Laying Out Relieving Cam. — As previously stated, the cams for relieving the teeth of formed cutters should be accurately made; a simple and efficient way is as follows: the cam-plate is first chucked out to fit the mandrel, and the sides and periphery turned in the usual way. The hole is then plugged up, as this space has to be utilized in laying out. This lay-out is illustrated in Fig. 27. First one actual sized cutter tooth is drawn or rather scribed, as shown, the object being

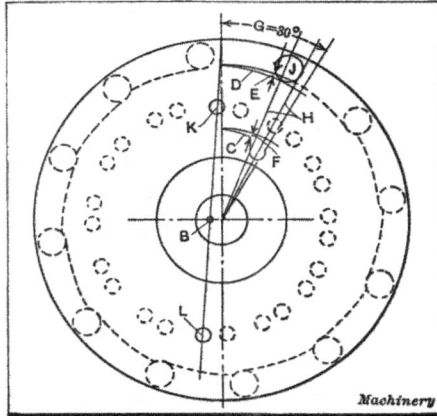

Fig. 27. Layout of One of the Relieving Cams

to determine the pitch of the clearance. It is necessary to have enough clearance to allow a free cutting action, but if too much is allowed, short-lived cutters having a tendency to chatter will result.

A good practical way to determine the correct amount of clearance is to use as a guide a formed cutter made by a reliable cutter manufacturer. Set this cutter central on the cam blank and carefully scribe the clearance line. This line is simply an arc set off center. With a pair of dividers, find the center and mark it with a center-punch mark. This center is shown at B, and the line of clearance at C. From the center B the arc D is scribed, which gives the same relation between the points E and F. The 30 degrees in the arc G represent one tooth. To make sure that the former will not start to return until the complete tooth has been relieved, and also that it will be ready to start cutting as soon as the next tooth comes around to the right position, the time of return is confined between the lines H. Next the circle J is laid off. This circle connects the lowest point of one stroke with the beginning of the next stroke. The center of this circle is carefully center-punched.

The two holes K and L are next laid off, and the centers punched. In this case there are twenty-four holes altogether, or two for each tooth. These holes are used for locating the cam while machining the rises.

Machining the Relieving Cam. — The first operation in machining the cam illustrated in Fig. 27 is that of boring the holes after they have been laid out as described in the preceding paragraph. The milling machine is used for this purpose and the spacing of the holes is effected by means of the dividing head. The head is set at right angles with the platen, its center being on a plane with the milling machine spindle. The cam blank is placed in position on the special mandrel having a washer and screw for holding it in position. The punch mark for the hole K is carefully centered, after which the hole is drilled, trued up with a boring tool and then reamed. The dividing head having previously been set for twelve divisions, the remainder of the holes corresponding to K are finished. Next the center for the hole L is located and all the holes in this circle finished. These holes are shown by the small dotted circles in Fig. 27. It is evident that these holes finished in the manner described will serve as an accurate means for locating the blank while machining the rises. Next the hole J is located, drilled, bored and reamed, and also the other holes in this outer circle. As this is comparatively heavy drilling to do in the manner described, the superfluous stock can be drilled on an upright drill, preparatory to boring and reaming on the milling machine.

The cam rises are next machined while the cam is held on an arbor provided with a flange containing two dowel pins spaced to match the holes K and L. It is evident that when the holes K and L in the cam blank are slipped over these dowels, the arc D is thrown in a position where it can be machined as part of a circle. This may be done in the shaper, the dividing head being used for this purpose. After planing one rise as shown by the dotted lines, the piece is moved into position for the next rise, by engaging the two dowel pins with the holes corresponding to that rise.

In the final finishing, all the rises are gone over one after the other with a very light cut and without changing the position of the tool. It is very necessary for all the rises to be alike; otherwise, cutters with uneven teeth would be the final result. For this reason, after the final finishing chip has been started it should be continued, for if the shaper should be allowed to remain idle for a time, some rises would surely be higher than others. Three of the holes can be used for doweling the cam to its holder, while three more can be countersunk and used for screw holes. The cam should be made of tool steel and it should be hardened to make it more durable.

Hobs for Worm-gears. — The hob for a worm-gear should not be made an exact duplicate of the worm. The corners at the top of the hob teeth should be rounded and fillets should also be provided at the root of the thread. The radii of these rounded corners and fillets should be as large as the clearance allows, or about one-twentieth of the circular pitch of the thread. The dimensions of the hob, which are different from the dimensions of the worm, are as follows:

Outside diameter of hob = outside diameter of worm + 0.1
   × linear pitch of worm.

Root diameter of hob = outside diameter of worm − 1.2732
   × linear pitch of worm.

If a hob is to be used a great deal and is subjected to much wear, it is advisable to increase the outside diameter above the dimensions given from 0.010 to 0.030 inch, according to its diameter and pitch, to allow for the decrease in diameter due to the relief and repeated grindings.

Dimensions for proportioning the tool for cutting the hob

thread are given in Fig. 28. When threading with such a tool, the depth of the thread is gaged by the shoulder on the tool, and accuracy is insured if the blank diameter is correct. The tool ought to be provided with side clearance of from 5 to 10 degrees from the angle of the thread. Grinding a tool of this kind, of course, changes its form so that it must not be used indefinitely in making a large number of similar hobs.

Number of Flutes or Gashes in Hobs. — For finding the approximate number of flutes in a hob, the following rule may be used: Multiply the diameter of the hob by 3, and divide this product by twice the linear pitch. This rule gives suitable results for hobs for general purposes; certain modifications, however, are often necessary. It is important that the number of flutes or gashes in hobs bear a certain relation to the number of threads in the hob and the number of teeth in the worm-wheel to be hobbed. In the first place, avoid having a common factor between the number of threads in the hob and the number of flutes; that is, if the worm is double-threaded, the number of gashes should be, say, 7 or 9, rather than 8. If it is triple-threaded, the number of gashes should be 7 or 11, rather

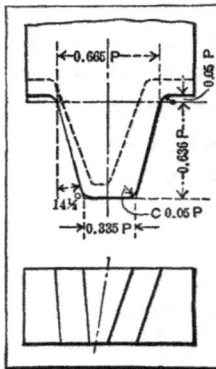

Fig. 28. Tool for Threading Worm-gear Hobs

than 6 or 9. The second requirement is to avoid having a common factor between the number of threads in the hob and the number of teeth in the worm-wheel. For example, if the number of teeth in the wheel is 28, it would be advisable to have the hob triple-threaded, as 3 is not a factor of 28. Again, if there were to be 36 teeth in the worm-gear, it would be preferable to have 5 threads in the hob.

The cutter used in gashing hobs should be from $\frac{1}{8}$ to $\frac{1}{4}$ inch thick at the periphery, according to the pitch of the hob thread. The width of the gash at the periphery of the hob should be about 0.4 times the pitch of the flutes. The cutter should be sunk into the blank so that it extends from $\frac{3}{16}$ to $\frac{1}{4}$ inch below the root of the thread.

Relieving Teeth of Spirally-fluted Hobs. — Hobs are generally fluted parallel with the axis, although the cutting action will be better if they are fluted on a helix or "spiral" at right angles with the thread. This is especially true if the lead of the hob is comparatively large. The difficulty of relieving the teeth with an ordinary backing-off or relieving attachment is often the cause of using a flute that is parallel with the axis. Flutes cut at right angles to the hob threads can, however, also be relieved with an ordinary attachment if the angle of the flute is slightly modified. In order to relieve hobs with a regular relieving attachment, it is necessary that the number of teeth in

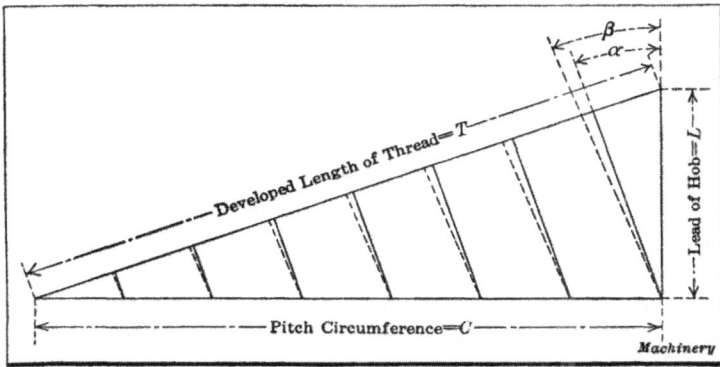

Fig. 29. Diagram Illustrating how Angle of Flute affects Spacing of Teeth along Thread Helix

one revolution along the thread, be such that the relieving attachment can be geared to suit this number. To obtain a number of teeth that can be relieved, the angle of the flute is changed so that the number of teeth in one complete turn of the thread will be either a whole number or at least some number for which the relieving attachment can be geared. It will be understood that this change is comparatively slight so that the flutes will be approximately at right angles to the thread. The effect which a comparatively slight change in the angle of the flute has upon the number of teeth per turn of the hob thread is illustrated by the diagram, Fig. 29. On this diagram, C represents the pitch circumference of the hob, and T equals the developed length of the thread. If we assume that there are seven flutes cut at an angle α so that they are exactly at right angles to the thread, there will be, in this particular case, about 7.8 teeth along a complete turn of the thread T. By changing this angle α slightly to angle β, as indicated by the dotted lines, one turn of the thread is divided into exactly eight flutes instead of 7.8 flutes.

The method of determining this angle β and the lead of the flute, to secure the required number of teeth along one turn of the thread, is indicated by the following example: Suppose we have a hob of 1.851 inch lead and 6.676 inches pitch circumference, that is to have twelve spiral flutes. What lead should the milling machine be geared for, and what flute angle will be required for an even number of teeth in one complete turn of the thread?

It is first necessary to find the number of teeth that there would be in a single turn of the hob thread if the flutes were milled exactly at right angles to the thread. This number equals the actual number of flutes in the hob divided by the square of the cosine of the thread helix angle measured from a line perpendicular to the hob axes. First obtain the thread helix angle. The tangent of this angle equals the lead of the thread divided by the circumference. Thus $1.851 \div 6.676 =$ 0.277 which is approximately the tangent of 15 degrees 30 minutes. The cosine of 15 degrees 30 minutes equals 0.9636; therefore the number of teeth in one turn of the thread, in this particular case, equals $12 \div 0.9636^2 = 12.92$ teeth.

An ordinary relieving attachment cannot be set for this number of teeth, but by making a slight change in the angle or lead of the spiral flute, we can obtain 13 teeth in one turn of the hob thread, instead of 12.92. The rule for finding the modified lead of the flute is as follows: Multiply the lead of the hob thread by the number of flutes in the hob, and divide the product by the difference between the actual number of flutes and the number of teeth required in a turn of the thread. Applying this rule, we have $\dfrac{1.851 \times 12}{13 - 12} = 22.21$ inches.

If the machine were geared for $22\frac{1}{4}$ inches, the result would be sufficiently accurate for practical purposes. If the flutes had been milled at right angles to the thread, the lead would have been 24.072 inches. To find the tangent of the angle at which the machine table should be set for a lead of $22\frac{1}{4}$ inches and a pitch circumference of 6.676 inches, divide the pitch circumference by the lead of the flute. Thus $6.676 \div 22.25 = 0.3$ which is the tangent of 16 degrees 42 minutes. By subtracting this modified fluting angle from 15 degrees 30 minutes, which

Factors for Determining Number of Teeth for Various Thread Angles

Thread angle, degrees.	Factor.	Thread angle, degrees.	Factor.	Thread angle, degrees.	Factor.	Thread angle, degrees.	Factor.	Thread angle, degrees.	Factor.
5	1.0076	27	1.2596	44	1.9326	49¾	2.3953	55	3.0396
6	1.0110	28	1.2827	44½	1.9657	50	2.4203	55¼	3.0779
7	1.0150	29	1.3073	45	2.0000	50¼	2.4457	55½	3.1170
8	1.0198	30	1.3333	45¼	2.0176	50½	2.4716	55¾	3.1571
9	1.0251	31	1.3612	45½	2.0356	50¾	2.4980	56	3.1980
10	1.0310	32	1.3905	45¾	2.0538	51	2.5250	56¼	3.2398
11	1.0378	33	1.4217	46	2.0723	51¼	2.5525	56½	3.2826
12	1.0452	34	1.4550	46¼	2.0912	51½	2.5806	56¾	3.3264
13	1.0533	35	1.4903	46½	2.1105	51¾	2.6091	57	3.3712
14	1.0622	36	1.5279	46¾	2.1300	52	2.6383	57¼	3.4170
15	1.0718	37	1.5678	47	2.1500	52¼	2.6680	57½	3.4639
16	1.0822	38	1.6104	47¼	2.1703	52½	2.6984	57¾	3.5119
17	1.0936	39	1.6558	47½	2.1910	52¾	2.7294	58	3.5611
18	1.1057	40	1.7041	47¾	2.2120	53	2.7611	58¼	3.6114
19	1.1186	40½	1.7295	48	2.2335	53¼	2.7934	58½	3.6629
20	1.1326	41	1.7557	48¼	2.2553	53½	2.8263	58¾	3.7147
21	1.1474	41½	1.7827	48½	2.2766	53¾	2.8600	59	3.7698
22	1.1632	42	1.8107	48¾	2.3002	54	2.8944	59¼	3.8253
23	1.1802	42½	1.8397	49	2.3233	54¼	2.9297	59½	3.8821
24	1.1982	43	1.8696	49¼	2.3469	54½	2.9655	59¾	3.9403
25	1.2174	43½	1.9006	49½	2.3709	54¾	3.0021	60	4.0000
26	1.2379	…	…….	…	…….	…	…….	…	…….

would be the angle if the flutes were milled at right angles to the thread, we obtain the amount that the flute varies from the right-angle position. In this case, the difference equals $(16° 42') - (15° 30') = 1° 12'$.

The number of teeth that would be formed in one turn of the hob thread can readily be determined by means of the accompanying table. In using this table, it is simply necessary to find the number corresponding to the thread angle and multiply it by the number of flutes in the hob. As the helix angle of the thread in this case is 15 degrees 30 minutes, the factor for this angle is found by taking the values between those given for 15 and 16 degrees. Thus, 1.0769 × 12 = 12.92 teeth in one complete turn of the thread.

The following rules cover the problems connected with the fluting of hobs (it is assumed that the angle of the thread is measured from a line perpendicular to the hob axis):

To find the angle of the thread helix: The tangent of the helix angle at the pitch circumference equals the lead of the hob thread divided by the pitch circumference.

To find the number of teeth in one turn of the hob thread: Divide the number of flutes by the square of the cosine of the helix angle of the thread.

To find the lead of the hob thread: Multiply the pitch circumference by the tangent of the helix angle of the thread.

To find the modified lead of the spiral flute: Multiply the lead of the hob thread by the number of flutes, and divide the product by the difference between the number of flutes and the number of teeth required in one turn of the hob thread.

To find the lead when the flute is square with the thread: Multiply the pitch circumference by the cotangent of the helix angle of the thread.

To find the angle between the flute and axis of hob: Divide the pitch circumference by the lead of the flute; the quotient equals the tangent of the angle.

Points on Making Reamers. — If reamers are to produce holes that are round and smooth, they must have cutting edges which are correctly spaced and formed. One common fault of reamers is that of chattering. This is generally due either to incorrect spacing of the cutting edges, excessive rake for the cutting edges, or too much clearance for the lands or tops of the teeth. When the teeth have too much rake, they cut too readily and produce a rough surface, whereas, excessive clearance for the lands causes chattering, as the reamer is not properly supported. The purpose for which the reamer is to be used must, of course, also be considered. Some of the principal points connected with reamer manufacture are given in the following, for the benefit of those who are required to make reamers occasionally on a small scale and with ordinary tool-room equipment.

Form and Angle of Reamer Flutes. — A strong tooth and flutes of sufficient depth to give room for chips are essential features of a reamer; both of these features depend on the number of teeth and the shape of the fluting cutter. In the smaller sizes there is no choice in regard to the number of teeth, as a fluted reamer with less than six teeth does not work well, and it should not have more than six. Straight reamers should have

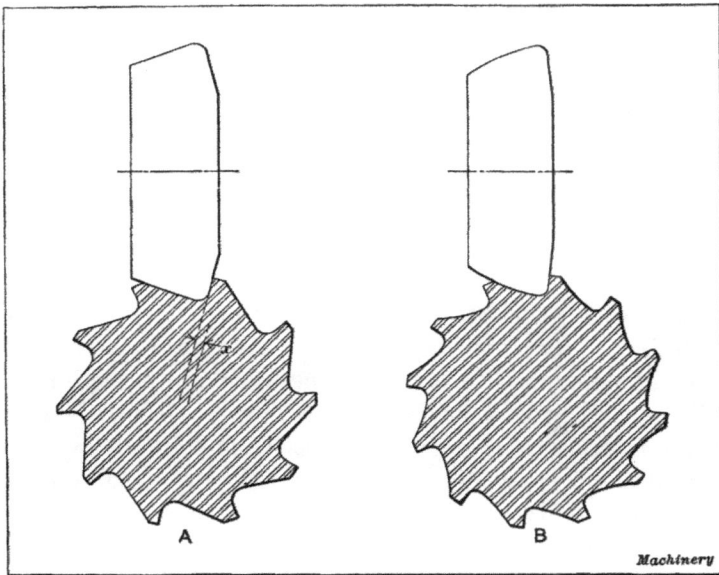

Fig. 30. Two Forms of Reamer Flutes

an even number of teeth so that the size of the reamer can be measured easily. Some of the so-called "formed" reamer fluting cutters produce a weaker tooth than an ordinary angular cutter. If a straight-fluted reamer is cut with an ordinary single-angle cutter having a vertical side, the cutter must be in good condition or a "ragged" tooth will result. A cutter combining several good features for the milling of straight-fluted reamers is shown at A, in Fig. 30. This cutter, by reason of the angular side, will cut a smooth tooth-face, and produce a strong tooth. A form of flute recommended by the Brown & Sharpe Mfg. Co. is shown at *B*. This form of flute provides ample clearance space and at the same time gives a strong tooth. The angle included between the cutting faces of the cutter shown at *A* is 85 degrees. This angle depends, of course, upon the angle desired for the reamer teeth, and it may be determined as follows: divide 360 by the number of teeth in the reamer and add the quotient to the desired reamer tooth angle.

Assuming a 1-inch reamer is to have eight 35-degree teeth, what should be the cutter angle?

$$360 \div 8 = 45; \quad 45 + 35 = 80;$$

hence, the fluting cutter should have an inclusive angle of 80 degrees.

The foregoing rule does not apply under all conditions. For instance, if we were to cut six 30-degree teeth in a $\frac{3}{4}$-inch reamer, we would, according to the formula, use a 90-degree cutter, but if we should use this same cutter for a $\frac{1}{2}$-inch reamer, the flutes would be rather shallow, and they would be still more so in the $\frac{3}{8}$-inch size. Cutters of $87\frac{1}{2}$- and 85-degree angles would be better adapted for these smaller sizes and they would form teeth having angles of $27\frac{1}{2}$ and 25 degrees, respectively. Ordinarily angles varying from 30 to 35 degrees will be found suitable for reamer teeth, although a 40-degree tooth is often more suitable for the larger sizes. This does not mean, however, that we can say off-hand what the tooth-angle shall be, and then proceed to cut any number of teeth on a certain size reamer, for, obviously, the number of teeth has a great deal to do with the angle of the teeth. In addition, the general appearance of the flute and the tooth depends largely on the rounded corners of the cutter used for fluting.

Position of Cutter for Fluting. — In general, reamers for steel and cast iron have the faces of the teeth either radial or slightly ahead of the center, but reamers for brass or bronze work better if the face of the tooth makes an angle of 5 or 6 degrees with the radial line, being that amount ahead of the center. The diagram

A, Fig. 30, illustrates how negative rake is obtained by setting the cutter slightly ahead of the center. The amount of this offset *x* may be obtained by dividing the reamer diameter by the constant 23; thus for a diameter of 1.5 inch the offset would equal 1.5 ÷ 23 = 0.065 inch. Hand reamers are ordinarily given negative rake, as they cut more smoothly than when the teeth are radial.

Number of Cutting Edges. — Fluted reamers for hand work rarely have less than six cutting edges. Even the smallest sizes should have six flutes if good results are expected, and the number should be increased in accordance with the diameter, as follows:

Reamer diameter............	⅛ to ½	1¹⁷⁄₃₂ to 1¼	1⁹⁄₃₂ to 1¾	1²⁵⁄₃₂ to 2¼	2⁹⁄₃₂ to 2¾
No. of flutes..........	6	8	10	12	14

Some mechanics claim that a reamer must have an odd number of teeth in order to work satisfactorily. Experience has shown, however, that an odd number of teeth is not essential to smooth accurate work; on the other hand, an even number is preferable owing to the convenience in measuring the size.

Irregular Spacing of Reamer Cutting Edges. — The spacing of the flutes of reamers is a problem concerning which there is a difference of opinion among toolmakers and mechanics generally. Some claim that if the flutes of half of the reamer are spaced irregularly but are made to correspond with the other half of the reamer, opposite cutting edges being exactly diametrically opposite, the object sought, *i.e.*, the elimination of chatter and the possibility of reaming a round hole, will be obtained. Others maintain that the cutting edges around the whole reamer should be irregularly spaced so that no two cutting edges are diametrically opposite. The advantages obtained by having the two halves of the reamer identical are that the reamer can be exactly measured, and that an equal width of the land of all the cutting edges can be more easily obtained if desired, as the milling machine table on which the reamer is mounted while fluting would have to be raised or lowered only half the number of times for obtaining this result, than would be the case if every tooth were irregularly spaced. It is the opinion of experienced reamer makers, however, that these advantages do not outweigh the disadvantages resulting from this method and that when the flutes are not irregularly spaced around the whole reamer, the tool is liable to chatter; moreover, if it once starts to cut a "cornered" hole, it will continue to do so. When all the cutting edges are diametrically opposite each other, the positions of each opposite pair coincide after half a revolution, whereas, if all the cutting edges are irregularly spaced there is no coincidence of position of any two cutting edges until the reamer has been turned around a complete revolution.

The error in measuring a reamer when all the cutting edges are irregularly spaced, and when no two are diametrically opposite, is very slight, provided the variation in spacing is not excessive. The irregularity may be so small that the error in measuring will not exceed 0.0003 inch. It has been the experience of a toolmaker of both mechanical and commercial experience that this is rather an advantage, as the reamer, when new, will be a small amount oversize, providing a slight allowance for wear which is not too great even for a hand reamer. The difference in the widths of the lands of the reamer teeth, which is the inevitable result of "breaking up" the flutes if the table or the cutter is not raised or lowered between consecutive cuts so as to make up for the difference in spacing, ordinarily is not considered important. The only reason for lands of even width would be for the sake of appearance, as the unequal width of the lands in no way interferes with the efficiency of the reamer. The commercial difficulties in making the lands of equal width, however, are rather too great to warrant unnecessary expense on account of a matter which has no mechanical importance.

Indexing for Irregular Spacing of Reamer Teeth. — When the flutes of a reamer have unequal spacing which is alike on each half so that opposite cutting edges are in line, the indexing may be done conveniently by milling the flutes in pairs; that is, after milling a flute, the index-head is turned half a revolution and the corresponding flute on the opposite side of the reamer is cut. Then, after milling the adjacent flute, the index-head is again turned half a revolution and so on. If the depths of the flutes are to be varied to secure lands of equal width, milling the flutes in pairs saves time as the cutter is only set once for opposite flutes. The flutes the first time around should be milled somewhat less than the required depth; then when milling them the second time, the cutter can be sunk to depth by noting the width of the land. After finishing one pair of opposite flutes, the cutter is re-set for the adjoining pair, etc.

To illustrate how the indexing movements for irregular spacing are determined, suppose a reamer is to have eight flutes with the

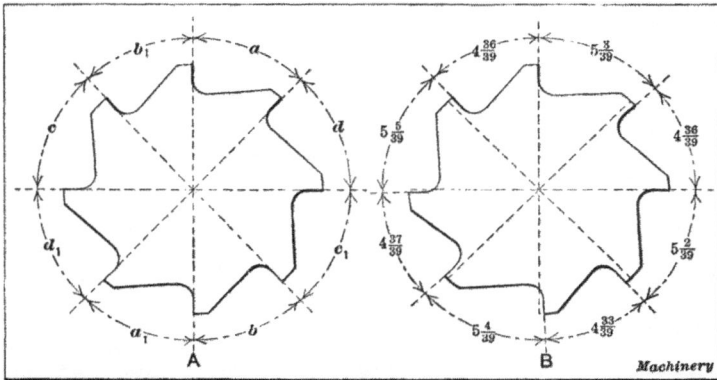

Fig. 31. (A) Irregular Spacing with each Half Uniform and Cutting Edges Opposite. (B) Flutes so Spaced that Cutting Edges are not Exactly Opposite

spacing of each half corresponding. Assuming that the 20-hole circle in the index plate is to be used, a complete revolution expressed as a total number of holes equals $20 \times 40 = 800$ holes. The number for 8 equal divisions equals $800 \div 8 = 100$ holes. The next thing to decide is the amount of irregularity in the spacing. The difference should be slight and need not exceed 2 degrees, although it is often made 3 or 4 degrees. Assuming that it is to be 2 degrees, the movement of the index crank necessary to give this variation must be determined. As 800 holes represent a complete revolution or 360 degrees, a movement of one hole equals

$\frac{360}{800}$, or nearly $\frac{1}{2}$ degree; therefore, the number of holes required for a movement of 2 degrees equals 2 ÷ 0.5 = 4 holes. Now if the divisions were all to be equal we would move the index crank 100 holes or 5 turns, but by varying the movement 4 holes one way or the other as near as this can be arranged, an irregularity of approximately 2 degrees is obtained. Thus the successive movements could be 96, 100, 103 and 101 holes, or 4 turns 16 holes, 5 turns, 5 turns 3 holes, and 5 turns 1 hole, respectively. Referring to diagram A, Fig. 31, flutes a and a_1 would be milled first diametrically opposite; then by indexing 96 holes the work would be located for milling flute b; after milling b_1 on the opposite side,

Irregular Spacing of Teeth in Reamers

Number of flutes in reamer......	4	6	8	10	12	14
Index circle to use............	39	39	39	39	39	49
Before cutting...	Move index crank the number of holes below more or less than for regular spacing.					
2d flute.....	8 less	4 less	3 less	2 less	4 less	3 less
3d flute.....	4 more	5 more	5 more	3 more	4 more	2 more
4th flute....	6 less	7 less	2 less	5 less	1 less	2 less
5th flute....	....	6 more	4 more	2 more	3 more	4 more
6th flute....	....	5 less	6 less	2 less	4 less	1 less
7th flute....	....	....	2 more	3 more	4 more	3 more
8th flute....	....	....	3 less	2 less	3 less	2 less
9th flute....	....	....	....	5 more	2 more	1 more
10th flute....	....	....	....	1 less	2 less	3 less
11th flute....	....	....	....	....	3 more	3 more
12th flute....	....	....	....	....	4 less	2 less
13th flute....	....	....	....	....	......	2 more
14th flute....	....	....	....	....	......	3 less

another movement of 100 holes would locate flute c; then after milling c_1 a movement of 103 holes would locate d; finally after fluting d_1 the cutter could be aligned with flute a_1 by a movement of 101 holes. The maximum amount of spacing between adjacent flutes is that represented by the spacing of flutes a_1 and b and equals 101 − 96 or 5 holes which is equal to $2\frac{1}{4}$ degrees. When selecting the numbers of holes by which the indexing movements are to be varied, one should remember that the total sum of the numbers must equal one-half the number of holes representing a complete revolution, when each half of the reamer is spaced alike and indexed as just described; thus,

$$96 + 100 + 103 + 101 = 400.$$

When all the flutes are to have irregular spacing, the indexing movements may be obtained from the accompanying table "Irregular Spacing of Teeth in Reamers." To illustrate its application, suppose a reamer is to have 8 flutes. If the spacing were equal, 5 turns of the index crank would be required ($40 \div 8 = 5$), but, as the table shows, the indexing movement for the second flute should be 5 turns minus 3 holes; for the third flute, 5 turns plus 5 holes, etc. In other words, the indexing movement for the second flute is $4\frac{36}{39}$ turns; for the third flute $5\frac{5}{39}$ turns; for the fourth flute $4\frac{37}{39}$ turns, etc., as shown by diagram B, Fig. 31. After milling the eighth flute, the cutter could be aligned with flute No. 1 by indexing $5\frac{3}{39}$ turns, as the illustration shows. This last movement is equal to 40 (the total number of turns required for one revolution of the dividing head) minus the total number of turns for the 7 flutes. Of course, it will be understood that the irregularity in spacing can be obtained by variations in indexing, other than those given in the foregoing.

Grinding Reamer Teeth. — A regular cutter and reamer grinder should preferably be used for grinding reamer teeth. Another method is to use an electric grinder held in the toolpost of a lathe while the reamer is held between the centers. The most economical and best results, however, are obtained with a grinder specially built for this work. The sharpening or backing off of reamers in a lathe is a poor makeshift, and should never be resorted to unless no other means are at hand. Sometimes it happens that a large reamer must be sharpened that is either too long or too large in diameter for the capacity of a grinder. In such a case the electric grinder, used in a lathe, will answer the purpose. This backing off or grinding the clearance on the teeth is an important operation. The amount of clearance that a reamer tooth should have depends somewhat upon the class of reamer, and its size. A point well worth remembering is this: A suitable clearance for reamer teeth, expressed in degrees (the face of the tooth being radial) is not a fixed amount, but varies for different diameters — the smaller the reamer the greater should be the angle of clearance. Note also that if the cutter and reamer grinder is set to grind the right amount of clearance on a large reamer, a small reamer would not have suitable clearance at that setting. Too great a clearance angle is apt to cause the reamer to chatter, particularly when working in brass or bronze. A newly ground reamer has a greater tendency to chatter than one that has been used for some time. If there be too little clearance, the reamer will not cut freely, as the lands will rub against the walls of the hole. In many cases the outcome will be a rough hole, or the reamer may "get stuck," and something will break if there is sufficient driving power. An eccentric relief, that is, one where the land back of the cutting edge is convex, rather than flat, is used by one or two manufacturers, and is preferable for finishing reamers, as the reamer will hold its size longer. When hand reamers are used merely for removing stock, or simply for enlarging holes, the flat relief is better, because the reamer has a keener cutting edge. The width of the land of the cutting edges should be about $\frac{1}{32}$ inch for a $\frac{1}{4}$-inch, $\frac{1}{16}$ inch for a 1-inch, and $\frac{3}{32}$ inch for a 3-inch reamer.

A simple method to regulate the clearance on a straight reamer is by raising or lowering the tooth-rest that the tooth slides on in passing in front of the wheel. The tooth that is being ground should always slide on the tooth-rest, because of the irregularities otherwise introduced, due to the liability of uneven spacing of the reamer teeth as the result of warping in hardening. The clearance of a taper reamer must not be regulated by adjusting the tooth-rest. When sharpening or backing off a taper reamer, the rest should stand in the plane of the axis of the reamer, and the desired clearance should be obtained by raising or lowering the reamer, relative to the wheel center. If a tooth is ground against the face it has a keener and smoother edge than when ground in the opposite direction. In spite of this good feature, the beginner should never attempt to grind a reamer that way, since there are several disadvantages connected with this method, which more than outweigh the advantages gained. In the first place the tendency of the wheel is to draw it away from the tooth-rest. If this happens, it usually means damage to the reamer and the wheel.

Reamers should be from 0.010 to 0.025 inch oversize before hardening, according to size and length of reamer, or, in general, $\frac{1}{2}$-inch size, 0.010 inch oversize; 1-inch, 0.015; 2-inch, 0.020 inch. On shell reamers allow about 0.005 inch more than these figures on all sizes. The diameter of the shank of a straight hand reamer should be from 0.001 to 0.002 inch below the diameter of the reamer. That part of the shank which is squared should be turned smaller in diameter than the shank itself, so that, when applying a wrench, no burr may be raised which may mar the reamed hole if the reamer is passed clear through it.

The grinding wheel used for backing off or sharpening reamers should be of fair size. Small wheels produce a cut of too great a curvature. The wheel should be of medium grade, as soft wheels produce a rough cut, and the hard ones become easily glazed, which is apt to draw the temper.

The rest should be made of tool steel and hardened, and the corners rounded to insure smooth action. It should be a trifle wider than the wheel. When backing off spiral reamers, the rest is set over to correspond with the angle of the spiral. A better way, however, is to make a special rest for spiral work, and grind this rest to the angle of the spiral.

Lead or Taper at End of Straight Reamers. — After the reamer is backed off and brought down to size a "lead" or taper is commonly ground at the end. This lead may be of various lengths to suit conditions. It must be very short or entirely absent if the reamer is intended to ream down to the bottom of holes, but for reaming through holes good results ought to be obtained with a lead of about $\frac{1}{2}$ inch in length. The taper should be about 0.005 or 0.006 inch. The back end of the reamer is also slightly tapered to prevent the ends of the teeth from scratching when passing through the hole. Finally, the chamfered corners are backed off, which finishes the reamer. The corners may be rounded instead of chamfered, but chamfered corners are easier to grind even, as the grinding can be done on centers.

Instead of simply grinding a slight taper at the end, hand reamers should be provided with a short straight guide at the end to insure reaming a straight hole. In general, the diameter of this guide should be from 0.005 to 0.010 inch smaller than the standard size of the reamer for diameters up to 1 inch, and from 0.010 to 0.015 inch smaller, for diameters from 1 to 3 inches. At the upper end of the guide there is a small groove for clearance when grinding, and then a tapered portion extending from about $\frac{3}{8}$ to $\frac{5}{8}$ inch for the smaller and from $\frac{3}{4}$ to $1\frac{1}{4}$ inch for the larger sizes. Commercial hand reamers made for the market are not generally provided with this guide, although it is a very valuable feature.

Helical Cutting Edges for Reamers. — Much has been written against the practice of making right-hand reamers with right-hand spiral flutes because they will be "drawn in" by the spiral; that this fear is without foundation in the case of a positive feed, such as is used in automatic screw machines, is obvious, but in a chucking machine, where the reaming is done by hand-feed, the reamer would tend to feed in rapidly. This would not apply to an end reamer, however, any more than to a four-lipped core-drill. Many such end-cutting reamers with right-hand spiral flutes have been made and are said to give better satisfaction than those with left-hand spiral or straight flutes, as the chips pass out more freely. Left-hand spirals on right-hand reamers are more apt to force the chips ahead, which is a good feature when reaming holes passing clear through, in a vertical machine such as a drill press. It is doubtful if the advantages of helical flutes for straight reamers intended for general work, offset the extra expense involved in milling the flutes. Helical flutes for straight reamers, however, are recommended when the holes to be reamed have lateral or crosswise openings. Steep taper reamers sometimes have right-hand helical cutting edges in order to improve the cutting action, and finishing taper and formed reamers in some cases have left-hand helical flutes to give a shaving cut, and thus leave a smoother surface.

Taper Reamers. — The most common taper reamers are those used for standard taper pins and for the standard taper sockets — Morse, Brown & Sharpe, and Jarno. Reamers for taper sockets are usually made in sets of two, a roughing reamer and a finishing reamer. The roughing reamer is made smaller in diameter at the small end than the finishing reamer and is provided with a groove cut like a thread along the cutting edges, as indicated in Fig. 32. The purpose of this groove is to break up the chips. It should be about $\frac{1}{16}$ inch wide, from $\frac{1}{32}$ to $\frac{1}{16}$ inch deep, and have a lead of about $\frac{1}{2}$ inch; it should, preferably, be cut with an Acme threading tool or an oil-grooving tool. As to whether

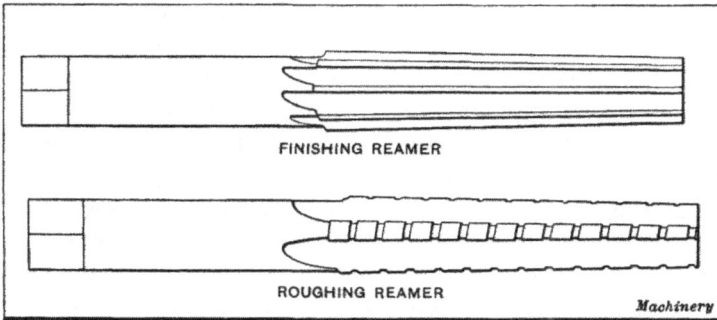

Fig. 32. Taper Roughing and Finishing Reamers

the groove should be right- or left-handed, there is a difference of opinion. Some toolmakers cut the groove left-handed on an ordinary right-hand reamer. Others claim, however, that this groove should be cut with a right-hand spiral on a right-hand reamer, because this gives the teeth a positive rake at the nicks along the flute, while a left-hand spiral gives a negative rake which causes trouble after the reamer has been in use for some time.

Roughing taper reamers are also made with right-handed spiral or helical cutting edges, especially if there is considerable taper, as the inclination of the cutting edges in the direction of the turning movement tends to draw the reamer in and gives a better cutting action. Ordinarily, however, the flutes of both roughing and finishing taper reamers are made straight. Finishing taper reamers are sometimes provided with left-hand helical flutes, the idea being that they will produce a smoother hole owing to the shaving cut.

Stepped Roughing Reamers. — Reamers for steep tapers are sometimes made in the form of a series of cylindrical steps. When making such a reamer, first turn the blank to the proper taper; then, using a round nose tool, cut a series of grooves about $\frac{1}{2}$ inch apart, $\frac{3}{32}$ inch deep and about $\frac{3}{32}$ inch wide, so that the reamer is nicked down a little at the end of each step. These dimensions may have to be modified according to the taper and size of the reamer. Next set the tailstock as for straight work, and turn each section or step straight. In this way a taper reamer that does not have a gradual increase in diameter, but which increases by small steps, is obtained. When using the reamer the teeth cut only at the ends of each step.

A very good roughing reamer may be made by cutting a right-hand spiral groove having about $\frac{3}{8}$- or $\frac{1}{2}$-inch lead, and then turning the lands between the convolutions of the groove straight, so that the blank is "stepped" spirally. This is done (leaving the lathe set for taper turning) with a square or broad-nosed tool, using the same feed as for cutting the groove. The edge of the tool must be set parallel with the axis of the work. Some toolmakers prefer left-hand spiral grooves for a reamer of this type. This reamer may be relieved towards the back when turning the blank by setting the broad-nosed tool so as to produce a slight back relief, instead of setting it exactly parallel with the axis. This would eliminate the necessity of backing off in the grinder, but the backed-off reamer gives better satisfaction.

Center Reamers. — Center reamers are made in two different styles. The older style has only one cutting edge formed by cutting away the metal down to the center of the tool and relieving the beveled portion of the remaining half so that a cutting edge is produced. The second and later style is provided with four or five flutes or cuts. These are straight and the lands between them are relieved on the beveled part. The included angle of the point is the same as that used for lathe centers, or 60 degrees. Use regular side milling cutters for fluting center reamers. Use a $2\frac{1}{2}$-inch cutter for a $\frac{1}{4}$-inch center reamer, and increase the cutter $\frac{1}{4}$ inch in diameter for each $\frac{1}{8}$-inch increase in center reamer size. This gives a 4-inch cutter for a 1-inch center reamer.

Center reamers for general use in the machine shop are commonly made from stock about $\frac{1}{2}$ or $\frac{5}{8}$ inch in diameter. The center reamer is more liable to chatter than some other reamers, which is quite troublesome in the case of arbors, etc., where a true center is wanted. A simple and quick way of obtaining a true center is to rough ream it with a four-tooth reamer, and then take a light shaving cut with a five-tooth reamer, using a slow speed. A center reamer for high-class work, such as arbors of which the centers have to be polished after hardening, should be in first-class condition; otherwise the center will present a mass of small ridges.

For heavy work, requiring larger centers, a reamer with a guide or teat gives good results; this guide is of the same size as the center drill (about $\frac{3}{16}$ or $\frac{1}{4}$ inch). Such a reamer is steadier in action, and cuts true with the center hole. A center reamer of this style, $\frac{1}{4}$ by $\frac{7}{8}$ inch diameter, should have five or six teeth.

Pipe Center Reamers. — By " center reamer " we generally understand a reamer, the teeth of which meet in a point so that very small centers may be reamed, but when large holes — usually cored — must be center-reamed, a large reamer is ordinarily used in which the teeth do not meet in a point. Center reamers for such work are called "bull" or "pipe" center reamers. These reamers should not be made for too large a range, as the teeth cannot be of good proportions at both ends of the reamer. When making tools of this type for reaming or chamfering holes from $\frac{1}{2}$ inch to 3 inches in diameter, one reamer should not be made to cover the whole range, but at least two should be provided; the smaller one should be suitable for holes from $\frac{1}{2}$ inch to about 2 inches, and the other for holes from about $1\frac{1}{2}$ inch to 3 inches, the former having 7 and the latter 11 teeth. A pipe center reamer should have an odd number of teeth, for it often happens that square holes, such as the blanks for socket wrenches, etc., have to be chamfered, and it can readily be seen that an odd number of teeth will give better results. Pipe center reamers used for square holes exclusively should have finer teeth than those for all-around service.

Rose Chucking Reamers. — These reamers are used for enlarging cored holes and are so constructed that they are able to remove a considerable amount of metal. The cutting edges are on a 45-degree bevel on the end of the reamer. The cylindrical part of the reamer has no cutting edges, but merely grooves cut for the full length of the reamer body, providing a way for the chips to escape and a channel for lubricant to reach the cutting edges. There is no relief on the cylindrical surface of the body part, but it is slightly back-tapered so that the diameter at the point with the beveled cutting edges is slightly larger than the diameter further back. The back-taper should not exceed 0.001 inch per inch. This form of reamer usually produces holes slightly larger than its size and it is, therefore, always made from 0.005 to 0.010 inch smaller than its nominal size, so that it may be followed by a fluted reamer for finishing.

Cutters for Fluting Rose Chucking Reamers. — The grooves on the cylindrical portion of rose chucking reamers are cut by a convex cutter having a width equal to from one-fifth to one-fourth the diameter of the rose reamer itself. The depth of the groove should be from one-eighth to one-sixth the diameter of the reamer. The width of the land of the cutting edges at the end should be about one-fifth the distance from tooth to tooth. If an angular cutter is preferred to a convex cutter for milling the grooves on the cylindrical portion, because of the higher cutting speed possible when milling, an 80-degree angular cutter slightly rounded at the point may be used.

The cutters used for fluting rose chucking reamers on the end are 80-degree angular cutters for $\frac{1}{4}$- and $\frac{5}{16}$-inch diameter reamers; 75-degree angular cutters for $\frac{3}{8}$- and $\frac{7}{16}$-inch reamers; and 70-degree angular cutters for all larger sizes. The grooves on the cylindrical portion are milled with convex cutters of approximately the following sizes for given diameters of reamers: $\frac{5}{32}$-inch convex cutter for $\frac{1}{2}$-inch reamers; $\frac{5}{16}$-inch cutter for 1-inch reamers; $\frac{3}{8}$-inch cutter for $1\frac{1}{2}$-inch reamers; $\frac{13}{32}$-inch cutters for 2-inch reamers; and $\frac{15}{32}$-inch cutters for $2\frac{1}{2}$-inch reamers. The smaller sizes of reamers, from $\frac{1}{4}$ to $\frac{3}{8}$ inch in diameter, are often milled with regular double-angle reamer fluting cutters having a radius of $\frac{1}{64}$ inch for $\frac{1}{4}$-inch reamer, and $\frac{1}{32}$ inch for $\frac{5}{16}$- and $\frac{3}{8}$-inch sizes.

Hand Taps. — Hand taps are ordinarily made in sets of three which are known as taper, plug, and bottoming taps. The point of the taper tap is turned down to the diameter at the bottom of the thread for a length of about three or four threads, and then about six threads are chamfered or tapered until the full diameter of the tap is reached. On the plug tap, three threads are chamfered at the point. On the bottoming tap only about one thread is chamfered. The diameter of the straight portion of the thread of all the taps is the same in the regular type of hand tap. Taps are made in sets, however, where only the bottoming tap is of the full diameter, while the other taps gradually decrease in diameter, so as to distribute the work between the three taps. A simple rule used by one tap manufacturer for proportioning U. S. standard thread taps when made in this manner, is to make the diameter of the first tap in a set equal to the diameter of the finishing tap less the depth of the thread, and to make the diameter of the second tap equal to the diameter of the finishing tap less one-third of the depth of the thread.

Proportions of Hand Taps. — The blanks for hand taps may be proportioned according to the following rules, for taps varying from three-sixteenths inch up to one inch in diameter:

Total length   = tap diameter $\times$ 3.5 + $1\frac{5}{8}$ inch;
Length of thread   = tap diameter + $1\frac{1}{2}$ inch;
Diameter of shank = root diameter of thread − 0.01 inch;
Size of square   = shank diameter $\times$ 0.75;
Length of square   = tap diameter $\times$ 0.75 + $\frac{1}{16}$ inch.

The proportions of hand taps for diameters larger than one inch should be as follows:

Total length   = tap diameter $\times$ 2.25 + $2\frac{7}{8}$ inches;
Length of thread   = tap diameter + $1\frac{1}{2}$ inch;
Diameter of shank = root diameter of thread − 0.02 inch;
Size of square   = shank diameter $\times$ 0.75;
Length of square   = tap diameter $\times$ 0.33 + $\frac{1}{2}$ inch.

Steel for Taps. — The best steel to use for tapping cast iron and brass is one containing from two to three per cent tungsten, but otherwise having the same composition as an ordinary high-carbon steel, that is, with from about 1.15 to 1.25 per cent carbon. This steel, if uniform in its composition, will contract or shorten about 0.002 inch per inch in hardening, the same as most carbon steels. When hardening, it should be heated to about 1525° F. The best steel for taps to be used on steel, as far as strength is concerned, is vanadium alloy steel containing from 0.25 up to 1 per cent of vanadium. The carbon content is the same as in regular carbon steels used for this purpose — from 1.15 to 1.25 per cent. The objection to this steel for taps is, however, that it is uncertain as regards its change in hardening. It is likely either to shorten or lengthen up to 0.002 inch per inch; but it is easily hardened and will stand a variation in the hardening heat of about 100° F. without injury to the tap. Expensive special steels are obtainable in the market that show practically no change in either the lead or the diameter of the tap when hardened, but these are not used commercially on account of their high cost. Good grades of English and Swedish steels are very uniform and are suitable for tap manufacture. They nearly always lengthen instead of shorten in hardening about 0.002 inch per inch. Some newer types of American steel show the same tendency. A high-speed steel suitable for taps should contain from 0.60 to 0.75 per cent carbon and from 15 to 20 per cent tungsten. This steel hardens at from 2100 to 2200° F. High-speed steel taps are especially good for automatic screw machine work, particularly when tapping brass or bronze.

The diameter of the stock from which the tap blank is to be turned should exceed the tap diameter by about $\frac{1}{16}$ inch for sizes up to $\frac{1}{2}$ inch, $\frac{1}{8}$ inch for sizes up to $1\frac{1}{2}$ inch, and $\frac{3}{16}$ to $\frac{1}{4}$ inch for larger diameters. This is to insure removing all of the outer de-carbonized surface. Before turning and threading the blank, anneal it by burying the heated steel in slaked lime or ashes.

Points on Tap Threading. — There are six important points to consider when threading a tap. The tap should have the correct diameter in the angle of the thread, a correct outside diameter, correct lead, correct angle between the sides of the thread, correct relation of this angle to the axis of the tap, and, finally, correct flats at the top and bottom of the threads, provided the thread is the U. S. standard form. The angle diameter, for instance, may be correct while the outside diameter would be a trifle large or small, depending upon whether the flat at the top of the thread were either too small or too large. The lead, of course, may be incorrect, while the other factors are practically correct. The angle of the thread may be larger or smaller than the standard angle, and if the lead, the outside diameter and the angle diameter were still approximately correct, the tap would produce a very poorly fitting thread. The angle between the sides of the thread may be correct in itself, but the thread-cutting tool may have been presented to the work at an oblique angle, thus producing a thread, the sides of which incline at different angles with relation to the axis of the tap. It is evident that all these requirements in regard to threading must be filled in order to make a perfect tap. The threading tool must, of course, be ground to the standard angle and the top of the tool should lie in a plane passing through the axis of the tap. When finishing, take very light cuts to secure a smooth thread, and use lard or sperm oil as a cutting lubricant. Some toolmakers use a single-pointed tool for roughing to within 0.002 or 0.003 inch of the size, and finish the thread with a chaser of the proper pitch.

Fluting Hand Taps. — The flutes of a tap serve two purposes. They form the cutting edges for the threads, and channels for the chips. The form of the flute is very important, as it affects the cutting qualities and strength of the tap. It is essential to have flutes that will cause the tap to work easily, but the strength of the tap should also be considered. In order to obtain strength, a shallow flute with no sharp corners is the first requirement, whereas an easy-working tap requires a considerable amount of chip room and, consequently, a comparatively deep flute. The correct form of flute, therefore, is a compromise between a flute which will give the greatest amount of chip room and the greatest strength to the tap.

Theoretically, the form of a tap flute should be varied in accordance with the material to be tapped.   Such refinement, however, is impracticable, and it is necessary to shape the flute so as to secure the best all-around results.   Three forms of flutes which are commonly used are shown at A, B and C, Fig. 33.   The form shown at A (which is obtained with a B. & S. cutter especially designed for grooving taps) gives a strong tap and one that is not liable to crack in hardening, because there are no sharp corners in the flute.   When fluting with this cutter the width w of the lands should be about two-tenths of the tap diameter.   Flute B is milled with an ordinary convex cutter.   This form is extensively used.   The depth d of the flute for a four-fluted tap is

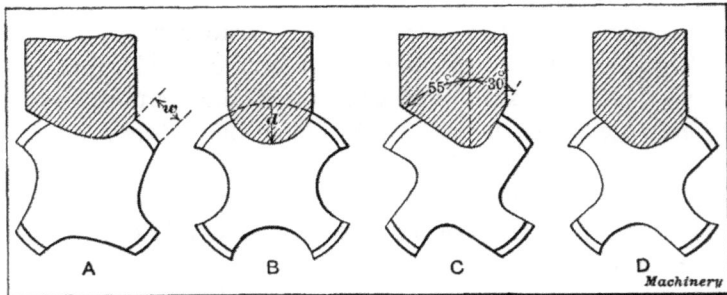

Fig. 33. Different Forms of Tap Flutes

made equal to the radius of the cutter; this radius equals one-fourth of the tap diameter.   The angular cutter C has an included angle of 85 degrees, 30 degrees on one side and 55 degrees on the other.   This cutter has been used to a great extent, but does not produce a very good form of flute.   The advantage of this cutter is that it is cheaper to use than formed cutters, and that one cutter can be used for a greater range of tap diameters.   The width of the land when using cutter C should equal one-fourth the tap diameter.   A tap fluted as at D has been found to give excellent results.   This is a comparatively new form of flute.   The depth of the flute is governed by the rule that the width of the land should equal one-half the width of the space, for four-fluted taps, or one-twelfth the circumference; for six-fluted taps the land should equal one-eighteenth of the circumference.

Taps are ordinarily fluted so that cutting edges are radial, unless they are to be used exclusively for brass, in which case a slight negative rake is sometimes given the teeth by locating the flute ahead of a radial line a distance varying from one-sixteenth to one-tenth the tap diameter.

In regard to the number of flutes there is some difference of opinion. There are those who consider four flutes the proper number to use on all sizes of taps; however, on large taps the land will be rather wide if made according to this rule, and better results will be obtained by increasing the number of flutes to six, for taps larger than $1\frac{3}{4}$ inch diameter.

Relief of Tap Threads. — When making hand taps the threads need not be relieved or "backed off," except on the chamfered or tapered portion, where the top of the thread should be slightly backed off. The diameter of the thread should, however, be made slightly smaller toward the shank than at the point. The amount of "back taper," as it is called, should be about 0.001 inch per inch. Some manufacturers relieve the thread in the angle for the full length of the tap, leaving the top of the thread cylindrical, and other makers leave one-third of the width of the land of the thread concentric or unrelieved both on the top of the thread and in the angle; the remaining two-thirds of the land are relieved both on the top and in the angle of the thread. In this way, the advantages of a good support for the tap in the hole to be threaded, and of a free and easy cutting tap, due to the relief, are obtained.

Taper Taps. — There are three important points to consider when making taper taps. In the first place, the threading tool must be presented to the tap at right angles to the axis of the tap, and not at right angles to its tapered surface, unless the tool is specially made for taper threading of taps with a definite taper. Second, taper taps should, if possible, be turned on lathes provided with taper attachments, and not by setting over the tailstock of the lathe, and, finally, proper relief should, in all cases, be given a taper tap. If a taper tap is not relieved either on the top or in the angle of the thread, it will not cut and if forced through a hole, will either leave a very rough and irregular thread, or break off its own teeth. It is imperative that taper taps be relieved the full length of the thread, on the top as well as in the angle of the thread, for the full width of the land. The relief should be greater on the leading side of the thread than on the rear side. This will lessen the friction and the resistance while cutting a thread. Tap manufacturers use a special machine for relieving taper taps; a three-cornered file is commonly used when special tools are not available.

Hob Taps. — Hob taps are intended only for the final finishing or sizing of the thread in dies. Straight hob taps are made with the same dimensions as regular hand taps. They are not relieved either on the top or in the angle of the thread of the straight portion, but two or at most three threads are chamfered at the point of the tap, and these chamfered threads are relieved on the top the same as are hand taps. A taper hob should be relieved slightly both on the top and in the angle of the thread for its whole length. The number of flutes is greater than in hand taps and should be as follows: $\frac{1}{4}$ to $\frac{7}{16}$ inch, six flutes; $\frac{1}{2}$ to $\frac{7}{8}$ inch, eight flutes; $1\frac{5}{8}$ to $1\frac{1}{2}$ inch, ten flutes; $1\frac{5}{8}$ to $2\frac{1}{2}$ inches, twelve flutes; $2\frac{5}{8}$ to 3 inches, fourteen flutes; $3\frac{1}{4}$ and larger, sixteen flutes. The flutes are cut with a single angle cutter with rounded point, the angle of the cutter being 50 degrees inclusive angle.

Milling Plate Cams. — Plate or peripheral cams having a constant rise may be milled by using the spiral head and a vertical spindle milling attachment on an ordinary milling machine. The cam blank is held in the spiral-head spindle and the spiral head is geared to the table feed-screw. An end mill is used and the cutting is done by the periphery or cylindrical part of the mill. The axis of the cutter spindle should be parallel to the axis of spiral-head spindle. It is evident that if the spiral head and cutter spindles are set in a vertical position, the lead of the cam or its rise in a complete revolution will equal the lead for which the machine is geared. If both spindles were at an angle, the lead of the cam would be less and, obviously, if they were in a horizontal position, the periphery of the cam would be milled concentric; hence, by setting the spindles at an angle, any rise that is less than the lead for which the machine is geared, can be obtained. To determine the angle at which the spiral head and cutter spindles should be set, first determine the hundredths of the cam circumference through which the rise is required; multiply the number of hundredths by the lead for which the milling machine is geared, and divide the rise of the cam by the product. The quotient is the sine of the required angle.

Example. — A cam is to be milled having a rise of 0.125 inch in 300 degrees or 0.83 of the cam circumference, and the machine is geared for a lead of 0.67 inch. At what angle should the spiral-head spindle be set?

$$0.83 \times 0.67 = 0.5561; \ 0.125 \div 0.5561 = 0.2247$$

which is approximately the sine of 13 degrees.

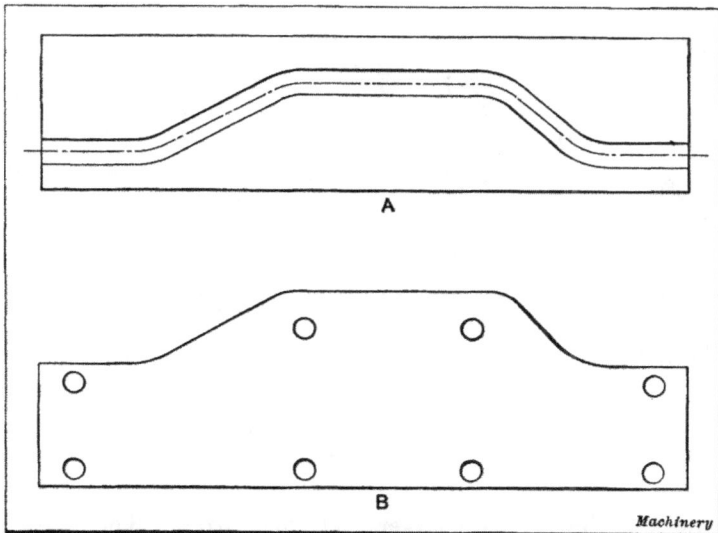

Fig. 34. (A) Development of Cam Circumference. (B) Templet used for Making Former

Making Master Cams. — The method of originating master, cylindrical cams, which is described in the following, has been used successfully in a shop where considerable of this work is done. A development of the cam at the surface of the cylinder is provided by the draftsman. If the cam is smaller than $2\frac{1}{2}$ or 3 inches diameter, or has unusually steep pitches, the development should, preferably, be laid out for a diameter two or three times larger than that of the desired cam.

Suppose it is desired to make a master for the cam, the development of which is shown at A, Fig. 34. The first step is to make a templet to match the development shown on the drawing. This templet may be made of mild steel, of a thickness depending upon the diameter to which it is to be bent. It may be fitted to the drawing with cold chisel and file, or, if considerable accuracy is desired, in the throw, the templet may be held in the milling machine vise, and the straight surfaces finished by milling. This templet, shown at B, in Fig. 34, is made for one side of the cam groove only.

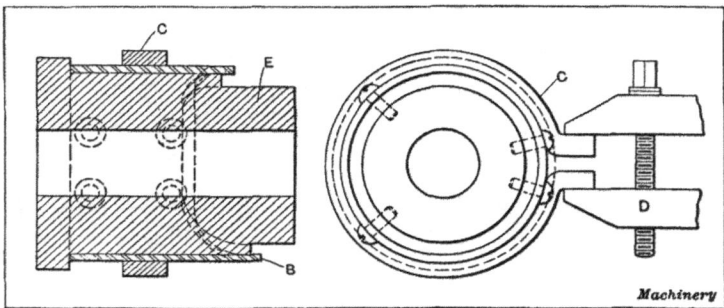

Fig. 35. Former which is used when Milling a Master Cam

The next step is to turn up a piece of steel or cast iron, as shown at E, Fig. 35, to such a diameter that when the templet B is wrapped around it, as shown, the ends will just barely meet. This diameter is about the thickness of the plate less than the diameter to which the development was laid out, but it should be left a little larger and then fitted. The plate is now clasped around the body, with the back edge pushed close up against the shoulder to insure proper alignment of the working surface of the cam. If any difficulty is experienced in this wrapping process, a circular strap may be bent up with projecting ends as shown at C; with the aid of a clamp D, the templet may be stretched around smoothly. The templet and the body may now be drilled and tapped for screws, as shown, and for dowels as well, if found necessary.

Scribe the contour of the cam onto the body E, remove the templet, place the body on an arbor in the index centers of the milling machine, and cut away the stock to a depth of about $\frac{1}{8}$ inch, $\frac{1}{16}$ inch back of the scribed line. This is for the purpose of providing a clearance underneath the working edge of the templet, as shown in Fig. 35. The templet may now be placed in position on the body, and fastened there. In this way a former is made for making a master cam.

Fig. 36 shows a milling machine arranged for cam cutting; E is a casting made to grip the finished face of the column, and

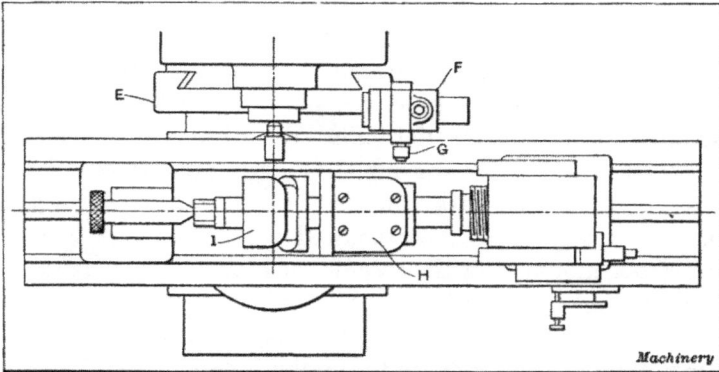

Fig. 36. Plan View of Milling Machine arranged for Milling Master Cam

carrying an adjustable block F; cam roll G is pivoted on a post which is adjustable in and out in block F; and the former H, and master cam blank I, are mounted, as shown, on an arbor in the index centers. By working the index worm crank, and the longitudinal feed crank together, roll G may be made to follow the outline of former H in such a fashion that the end mill will cut the desired groove in cam blank I. A slightly smaller mill may be used for a roughing cut, but, obviously, the roll and the finishing mill must be of about the same size if a true copy of the templet is desired. It will be found easier to follow the outline with the roll if the steeper curves are traced down rather than up.

A fairly good cam-cutting machine for making copies of the master cam I, may be improvised by using the attachment E,

F, G in a rack-feed machine. It might also be feasible to con-
nect the index worm with the telescopic feed shaft so as to give
a power feed to the contrivance. To insure accurate cams, the
arrangement for holding the tool must be made stiff enough to
move the table without much spring, or the table must be
weighted, so as to bring the pressure of the roll constantly
against one face of the master cam.

Graduating on the Milling Machine. — A milling machine
equipped with a spiral head is sometimes used for graduating
verniers, flat scales and other parts requiring odd fractional
divisions or graduations. The spiral-head spindle should be
geared to the table feed-screw so that a longitudinal movement
of the table is secured by turning the indexing crank. When
using a Brown & Sharpe machine, the gear for the spiral head
is mounted on the differential index center inserted in the main
spiral-head spindle. By varying the indexing movement, graduations can be spaced with considerable accuracy. The graduation lines are cut by a sharp-pointed tool held either in a fly-
cutter arbor mounted in the spindle, or between the collars of a
regular milling cutter arbor. The lines are drawn by feeding
the table laterally by hand, and the lengths of the lines representing different divisions and sub-divisions can be varied by
noting the graduations on the cross-feed screw. The gearing
between the spiral head spindle and feed-screw should be equal
or of such a ratio that the feed-screw and spindle rotate at the
same speed. Assuming that the lead of the feed-screw thread
is 0.25 inch and that 40 turns are required for one revolution of
the spiral-head spindle, then one turn of the index crank will
cause the table to move longitudinally a distance equal to one-
fortieth of 0.25 or 0.00625 inch. $\left(\frac{1}{40} \times \frac{25}{100} = 0.00625.\right)$ Suppose graduation lines 0.03125 or $\frac{1}{32}$ inch apart were required
on a scale. Then the number of turns of the index crank for
moving the table 0.03125 inch equals 0.03125 ÷ 0.00625 = 5
turns.

If the divisions on a vernier reading to thousandths of an inch
were to be 0.024 inch apart, the indexing movement would equal

Indexing Movements for Graduating on Milling Machine

Movement of table.	Holes.	Circle.	Movement of table.	Holes.	Circle.	Movement of table.	Holes.	Circle.
0.0001275	1	49	0.0006377	5	49	0.0011479	9	49
0.0001330	1	47	0.0006410	4	39	0.0011574	5	27
0.0001454	1	43	0.0006465	3	29	0.0011628	8	43
0.0001524	1	41	0.0006579	2	19	0.0011718	3	16
0.0001603	1	39	0.0006649	5	47	0.0011824	7	37
0.0001689	1	37	0.0006757	4	37	0.0011905	4	21
0.0001894	1	33	0.0006944	3	27	0.0011968	9	47
0.0002016	1	31	0.0006944	2	18	0.0012096	6	31
0.0002155	1	29	0.0007268	5	43	0.0012195	8	41
0.0002315	1	27	0.0007353	2	17	0.0012500	4	20
0.0002551	2	49	0.0007576	4	33	0.0012500	3	15
0.0002660	2	47	0.0007622	5	41	0.0012755	10	49
0.0002717	1	23	0.0007653	6	49	0.0012820	8	39
0.0002907	2	43	0.0007813	2	16	0.0012930	6	29
0.0002976	1	21	0.0007979	6	47	0.0013081	9	43
0.0003049	2	41	0.0008012	5	39	0.0013158	4	19
0.0003125	1	20	0.0008064	4	31	0.0013257	7	33
0.0003205	2	39	0.0008152	3	23	0.0013298	10	47
0.0003289	1	19	0.0008333	2	15	0.0013513	8	37
0.0003378	2	37	0.0008446	5	37	0.0013587	5	23
0.0003472	1	18	0.0008621	4	29	0.0013722	9	41
0.0003676	1	17	0.0008721	6	43	0.0013888	6	27
0.0003788	2	33	0.0008929	7	49	0.0013888	4	18
0.0003826	3	49	0.0008929	3	21	0.0014031	11	49
0.0003906	1	16	0.0009146	6	41	0.0014113	7	31
0.0003989	3	47	0.0009259	4	27	0.0014422	9	39
0.0004032	2	31	0.0009308	7	47	0.0014535	10	43
0.0004167	1	15	0.0009375	3	20	0.0014628	11	47
0.0004310	2	29	0.0009469	5	33	0.0014706	4	17
0.0004361	3	43	0.0009616	6	39	0.0014881	5	21
0.0004573	3	41	0.0009869	3	19	0.0015086	7	29
0.0004630	2	27	0.0010081	5	31	0.0015152	8	33
0.0004808	3	39	0.0010136	6	37	0.0015202	9	37
0.0005068	3	37	0.0010174	7	43	0.0015244	10	41
0.0005102	4	49	0.0010204	8	49	0.0015306	12	49
0.0005319	4	47	0.0010417	3	18	0.0015625	5	20
0.0005435	2	23	0.0010638	8	47	0.0015625	4	16
0.0005682	3	33	0.0010671	7	41	0.0015957	12	47
0.0005814	4	43	0.0010776	5	29	0.0015989	11	43
0.0005952	2	41	0.0010869	4	23	0.0016026	10	39
0.0006048	3	31	0.0011029	3	17	0.0016128	8	31
0.0006098	4	41	0.0011218	7	39	0.0016204	7	27
0.0006250	2	20	0.0011363	6	33	0.0016303	6	23

Indexing Movements for Graduating on Milling Machine — *Continued*

Movement of table.	Holes.	Circle.	Movement of table.	Holes.	Circle.	Movement of table.	Holes.	Circle.
0.0016447	5	19	0.0021682	17	49	0.0026785	9	21
0.0016581	13	49	0.0021738	8	23	0.0026785	21	49
0.0016666	4	15	0.0021802	15	43	0.0027028	16	37
0.0016768	11	41	0.0021875	7	20	0.0027174	10	23
0.0016892	10	37	0.0021960	13	37	0.0027243	17	39
0.0017045	9	33	0.0022059	6	17	0.0027344	7	16
0.0017241	8	29	0.0022176	11	31	0.0027440	18	41
0.0017288	13	47	0.0022436	14	39	0.0027618	19	43
0.0017361	5	18	0.0022607	17	47	0.0027777	8	18
0.0017442	12	43	0.0022728	12	33	0.0027777	12	27
0.0017628	11	39	0.0022866	15	41	0.0027925	21	47
0.0017857	6	21	0.0022959	18	49	0.0028017	13	29
0.0017857	14	49	0.0023027	7	19	0.0028060	22	49
0.0018144	9	31	0.0023148	10	27	0.0028125	9	20
0.0018292	12	41	0.0023257	16	43	0.0028225	14	31
0.0018382	5	17	0.0023438	6	16	0.0028409	15	33
0.0018518	8	27	0.0023649	14	37	0.0028717	17	37
0.0018581	11	37	0.0023706	11	29	0.0028846	18	39
0.0018617	14	47	0.0023809	8	21	0.0028963	19	41
0.0018750	6	20	0.0023937	18	47	0.0029070	20	43
0.0018896	13	43	0.0024038	15	39	0.0029167	7	15
0.0018939	10	33	0.0024192	12	31	0.0029256	22	47
0.0019021	7	23	0.0024235	19	49	0.0029337	23	49
0.0019132	15	49	0.0024306	7	18	0.0029412	8	17
0.0019231	12	39	0.0024390	16	41	0.0029605	9	19
0.0019396	9	29	0.0024455	9	23	0.0029762	10	21
0.0019532	5	16	0.0024622	13	33	0.0029890	11	23
0.0019737	6	19	0.0024710	17	43	0.0030094	13	27
0.0019818	13	41	0.0025000	8	20	0.0030172	14	29
0.0019947	15	47	0.0025000	6	15	0.0030241	15	31
0.0020161	10	31	0.0025266	19	47	0.0030303	16	33
0.0020271	12	37	0.0025339	15	37	0.0030406	18	37
0.0020350	14	43	0.0025463	11	27	0.0030448	19	39
0.0020485	16	49	0.0025510	20	49	0.0030488	20	41
0.0020833	13	39	0.0025640	16	39	0.0030524	21	43
0.0020833	5	15	0.0025736	7	17	0.0030586	23	47
0.0020833	11	33	0.0025862	12	29	0.0030611	24	49
0.0020833	9	27	0.0025915	17	41	0.0031250	9	18
0.0020833	7	21	0.0026164	18	43	0.0031250	10	20
0.0020833	6	18	0.0026209	13	31	0.0031250	8	16
0.0021277	16	47	0.0026316	8	19	0.0031889	25	49
0.0021342	14	41	0.0026515	14	33	0.0031915	24	47
0.0021552	10	29	0.0026596	20	47	0.0031978	22	43

Indexing Movements for Graduating on Milling Machine — *Continued*

Movement of table.	Holes.	Circle.	Movement of table.	Holes.	Circle.	Movement of table.	Holes.	Circle.
0.0032014	21	41	0.0036990	29	49	0.0042091	33	49
0.0032050	20	39	0.0037038	16	27	0.0042152	29	43
0.0032095	19	37	0.0037163	22	37	0.0042232	25	37
0.0032197	17	33	0.0037234	28	47	0.0042338	21	31
0.0032257	16	31	0.0037500	12	20	0.0042553	32	47
0.0032327	15	29	0.0037500	9	15	0.0042685	28	41
0.0032408	14	27	0.0037793	26	43	0.0042765	13	19
0.0032607	12	23	0.0037878	20	33	0.0042971	11	16
0.0032738	11	21	0.0038043	14	23	0.0043104	20	29
0.0032895	10	19	0.0038112	25	41	0.0043268	27	39
0.0033088	9	17	0.0038195	11	18	0.0043368	34	49
0.0033164	26	49	0.0038265	30	49	0.0043477	16	23
0.0033245	25	47	0.0038305	19	31	0.0043562	23	33
0.0033333	8	15	0.0038460	24	39	0.0043605	30	43
0.0033431	23	43	0.0038564	29	47	0.0043750	14	20
0.0033538	22	41	0.0038692	13	21	0.0043883	33	47
0.0033654	21	39	0.0038794	18	29	0.0043922	26	37
0.0033784	20	37	0.0038853	23	37	0.0043980	19	27
0.0034091	18	33	0.0039063	10	16	0.0044119	12	17
0.0034273	17	31	0.0039246	27	43	0.0044210	29	41
0.0034375	11	20	0.0039352	17	27	0.0044354	22	31
0.0034439	27	49	0.0039475	12	19	0.0044643	15	21
0.0034482	16	29	0.0039540	31	49	0.0044643	35	49
0.0034574	26	47	0.0039636	26	41	0.0044871	28	39
0.0034722	10	18	0.0039773	21	33	0.0045060	31	43
0.0034722	15	27	0.0039894	30	47	0.0045140	13	18
0.0034885	24	43	0.0040064	25	39	0.0045213	34	47
0.0035063	23	41	0.0040322	20	31	0.0045259	21	29
0.0035156	9	16	0.0040443	11	17	0.0045452	24	33
0.0035255	22	39	0.0040541	24	37	0.0045610	27	37
0.0035325	13	23	0.0040625	13	20	0.0045732	30	41
0.0035474	21	37	0.0040700	28	43	0.0045835	11	15
0.0035714	12	21	0.0040759	15	23	0.0045920	36	49
0.0035714	28	49	0.0040817	32	49	0.0046055	14	19
0.0035904	27	47	0.0040948	19	29	0.0046194	17	23
0.0035984	19	33	0.0041160	27	41	0.0046296	20	27
0.0036186	11	19	0.0041223	31	47	0.0046371	23	31
0.0036289	18	31	0.0041666	22	33	0.0046473	29	39
0.0036339	25	43	0.0041666	14	21	0.0046512	32	43
0.0036585	24	41	0.0041666	18	27	0.0046543	35	47
0.0036637	17	29	0.0041666	12	18	0.0046875	15	20
0.0036765	10	17	0.0041666	10	15	0.0046875	12	16
0.0036858	23	39	0.0041666	26	39	0.0047195	37	49

Indexing Movements for Graduating on Milling Machine — *Continued.*

Movement of table.	Holes.	Circle.	Movement of table.	Holes.	Circle.	Movement of table.	Holes.	Circle.
0.0047256	31	41	0.0052296	41	49	0.0057180	43	47
0.0047299	28	37	0.0052327	36	43	0.0057400	45	49
0.0047349	25	33	0.0052365	31	37	0.0057433	34	37
0.0047414	22	29	0.0052419	26	31	0.0057692	36	39
0.0047620	16	21	0.0052635	16	19	0.0057874	25	27
0.0047796	13	17	0.0052884	33	39	0.0057927	38	41
0.0047873	36	47	0.0053030	28	33	0.0058142	40	43
0.0047968	33	43	0.0053125	17	20	0.0058187	27	29
0.0048074	30	39	0.0053194	40	47	0.0058336	14	15
0.0048384	24	31	0.0053242	23	27	0.0058466	29	31
0.0048470	38	49	0.0053364	35	41	0.0058512	44	47
0.0048613	14	18	0.0053572	42	49	0.0058599	15	16
0.0048613	21	27	0.0053572	18	21	0.0058674	46	49
0.0048782	32	41	0.0053781	37	43	0.0058710	31	33
0.0048912	18	23	0.0053880	25	29	0.0058825	16	17
0.0048989	29	37	0.0054057	32	37	0.0059027	17	18
0.0049202	37	47	0.0054170	13	15	0.0059122	35	37
0.0049244	26	33	0.0054348	20	23	0.0059215	18	19
0.0049345	15	19	0.0054434	27	31	0.0059294	37	39
0.0049420	34	43	0.0054486	34	39	0.0059375	19	20
0.0049569	23	29	0.0054522	41	47	0.0059455	39	41
0.0049677	31	39	0.0054690	14	16	0.0059524	20	21
0.0049745	39	49	0.0054848	43	49	0.0059598	41	43
0.0050000	16	20	0.0054878	36	41	0.0059782	22	23
0.0050000	12	15	0.0054924	29	33	0.0059841	45	47
0.0050308	33	41	0.0055148	15	17	0.0059951	47	49
0.0050402	25	31	0.0055238	38	43	0.0060188	26	27
0.0050532	38	47	0.0055555	24	27	0.0060346	28	29
0.0050596	17	21	0.0055555	16	18	0.0060480	30	31
0.0050676	30	37	0.0055746	33	37	0.0060607	32	33
0.0050785	13	16	0.0055852	42	47	0.0060812	36	37
0.0050876	35	43	0.0055925	17	19	0.0060898	38	39
0.0050928	22	27	0.0056035	26	29	0.0060980	40	41
0.0051022	40	49	0.0056088	35	39	0.0061052	42	43
0.0051136	27	33	0.0056123	44	49	0.0061171	46	47
0.0051281	32	39	0.0056250	18	20	0.0061224	48	49
0.0051474	14	17	0.0056403	37	41	0.0062500	...	1
0.0051627	19	23	0.0056450	28	31	....	...	..
0.0051721	24	29	0.0056546	19	21	....	...	..
0.0051830	34	41	0.0056690	39	43	....	...	..
0.0051861	39	47	0.0056816	30	33	....	...	..
0.0052083	15	18	0.0057065	21	23	....	...	..

0.024 ÷ 0.00625 = 3.84 turns. This fractional movement of 0.84 turn can be obtained within very close limits by indexing 26 holes in the 31-hole circle; thus, three complete turns will move the work 0.00625 × 3 = 0.01875 inch and $\frac{26}{31}$ turn will give a longitudinal movement equal to 0.00524+; therefore a movement of $3\frac{26}{31}$ turns = 0.01875 + 0.00524 = 0.02399 inch which is 0.00001 inch less than the required amount.

The accompanying table " Indexing Movements for Graduating on Milling Machine " was compiled by the Brown & Sharpe Mfg. Co., and will be found very convenient for determining what circle to use and the number of holes to index for given dimensions. The whole number of turns required is first determined and then the indexing movement for the remaining distance is taken from the table. Thus, for graduation lines 0.0218 inch apart, three complete turns give a movement of 0.00625 × 3 = 0.01875 inch. 0.0218 − 0.01875 = 0.00305 inch. By referring to the table it will be seen that a movement of 21 holes in the 43-hole circle equals 0.00305+. Therefore, to index the work 0.0218 inch, the crank should be given $3\frac{21}{43}$ turns. In graduating in this way, the index crank should always be turned in one direction after beginning the graduating operation, in order to prevent errors from any play or backlash that might exist between the table feed-screw and nut.

Diamond Tools. — In certain classes of work where great accuracy and precision are primary requirements, or when extremely fine lines are essential, the diamond is the only material that answers the purpose. There are also some materials which, because of their hardness, structure, or non-conductivity of heat, cannot be worked economically by means of steel tools. The latter become worn rapidly, losing their shape and dimensions to such a degree that the work produced is inaccurate, thus causing constant interruption of operation, loss of time and the use of new tools or frequent re-grinding or shaping of the old ones. This causes great expense and delay in production. Hard rubber, paper, and hardened steel cannot readily be worked by the use of steel tools, as is also the case with hard stone. As is well known, the diamond tool is indispensable for truing grinding wheels, whereas diamond dust is commonly used for lapping or grinding small precision work in tool-rooms, watch factories, etc., where great accuracy is required.

The diamonds which are used for grinding, turning, truing grinding wheels, etc., are of two kinds which are entirely different in appearance and quality. First, there is the black diamond or "carbon" which has a very dark purple-brown color and is an amorphous, granular stone with rarely any crystallization visible. The black diamond is the hardest material known and has considerable strength.

Second, there is the bort, which is entirely crystalline and is generally transparent and of all colors of the rainbow, as well as clear as glass. The latter is considered of greater hardness than all other bort, except that which is almost black. Bort is extremely brittle and is readily fractured or "cleaved" in the three directions of its cleavage planes parallel to the sides of the octahedral crystal, in which shape it is most commonly found.

Black diamonds or carbons have been quite extensively used by the National Meter Works in the production of some of the working parts of water meters which are made of hard rubber. It is necessary to machine these rubber parts after they are molded, and this substance is one of the most difficult to machine with steel tools; in fact, it is impossible to work it accurately with ordinary steel tools. A most peculiar property of hard rubber developed by the combination of ingredients necessary for vulcanizing, causes it to blunt the edge of steel tools so quickly that it is rarely possible to go over the surface of a small piece and have it pass the inspectors' limit gages, the wear of the tool being so great; consequently, diamond tools are used for turning this rubber.

Diamond tools have also been used to some extent for turning and boring operations in connection with machine-shop work. The most serious draw-back to the use of diamond tools is, of course, their cost. It has also been found difficult to hold an irregular shaped diamond so as to present the cutting edge advantageously and still hold it firmly enough to prevent its being dragged out by the pressure of the cut, and lost. Ordinarily, however, this is not a very serious objection as the diamond tool is simply used for very light finishing cuts. Another and more serious trouble is that of grinding the diamond to the correct shape for cutting.

Another very important use of the diamond is in the production of dies for drawing fine wire. Formerly, all small wire was drawn through holes in hardened steel plates, but these wear so rapidly that the wire soon loses its size and becomes inaccurate. A diamond die used for drawing wire consists of a bronze block in which a diamond is set, having a tapering polished hole through which the wire is drawn. The holes in these diamond dies are rarely any larger than 0.064 inch diameter, because steel plates or dies are considered sufficiently accurate and economical for larger sizes; also because of the greater cost of diamond dies. In order to show why such expensive material as diamonds can be used economically for this work, it may be stated that diamond dies may be used constantly for eight years and one die of 0.004 inch size, according to the record, has drawn over 550,000 pounds of soft copper wire.

Comparatively high temperatures do not affect the diamonds either in their hardness or, when sound, in their solidity, and does not produce checks or other flaws. A temperature higher than that sufficient to melt steel will, however, burn the diamond and that of the electric arc will do so readily. The diamond tools used for accurate work are all shaped by cutting and polishing so as to imitate the customary shape of steel tools. While bort is not as hard as a black diamond, it is considerably lower in price. For truing soft grinding wheels, bort may be more economical than the more expensive stone, but, as a general rule, the black diamond is cheaper in the end.

CHAPTER VII

PRECISION BENCH LATHE PRACTICE

The modern bench lathe finds wide application in the manufacture of small parts requiring considerable accuracy, as well as in fine tool work, where its facility of operation and its accuracy make it an ideal tool. Bench lathes have been developed to the same high standard of efficiency as the heavier types of lathes, and the design of various attachments has broadened the field of these machines so that they are able to handle a wide range of work. In addition to their adaptation to precision turning and boring operations, bench lathes may be equipped with attachments for milling and grinding, for chasing, cutting, and milling screw threads, for turret work, filing, and a variety of other operations. Many of these attachments, such as those for milling, grinding, threading, etc., are standard equipment supplied by bench lathe manufacturers, but many special attachments are also used in connection with bench lathe practice. All the parts and attachments should, as far as possible, be interchangeable, so that when there are several lathes, a given attachment can be used on any machine on which it may be required.

Experience has shown that a proper countershaft equipment adds greatly to the efficiency of any bench lathe. It is well to fasten the countershaft to a pipe frame attached to the bench and ceiling. This frame may be made of ordinary gas pipe, threaded at the ends for feet that are bolted to the bench and ceiling. If the countershafts used with these frames are designed with sockets which fit the pipe frame, so that they can be adjusted to any desired point, this makes it possible to use an endless belt and to adjust the belt tension by raising or lowering the countershaft on the vertical supports.

The general class of work for which the bench lathe is adapted and the use of different tools and attachments, are illustrated by the following examples from actual practice.

Precision Jig Work in the Bench Lathe. — In the manufacture of fine machinery, and for the finer work of the tool-room, the bench lathe is found invaluable, as the ordinary engine lathe is too large and unwieldy for this class of work. As an example of the kind of work for which the bench lathe is adapted, we will describe in detail the way in which two small jigs are made,

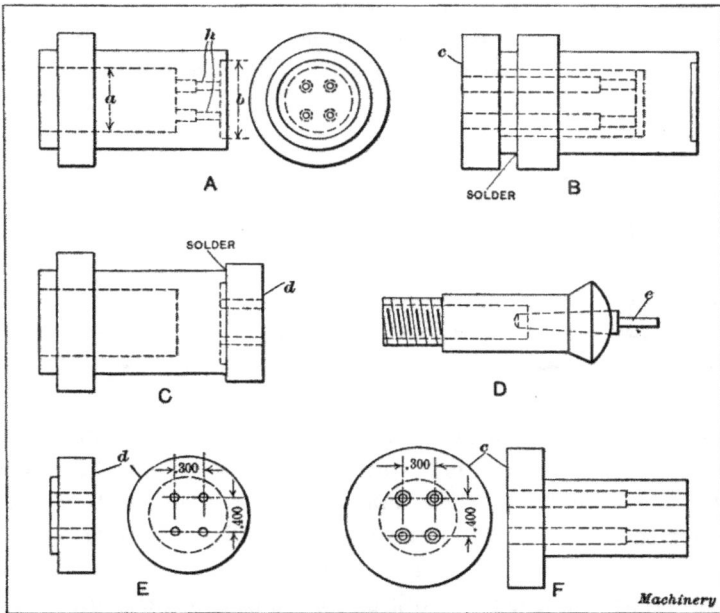

Fig. 1. **Examples of Precision Bench Lathe Work**

which are to be used for drilling the holes h in the die shown at A, Fig. 1. As the inner half of each of these holes is enlarged, the drilling is done from each side of the die, the jigs c and d being inserted as shown at B and C, respectively. Jig c is used for the larger holes, and d for the smaller ones. These jigs, which are shown in detail at E and F, are made of brass. While they are being turned, they are held in a universal chuck screwed on the bench lathe spindle. The shoulder on jig d is fitted accurately to the recess b in the die, and the body of jig c is also turned to a close fit for the hole a.

In order to bore the holes in each of these jigs in exactly the same relative positions, a steel master-plate A, Fig. 2, is first made, the use of which will be explained presently. This plate should first be turned in an engine lathe; it is then strapped to the faceplate F of the bench lathe (Fig. 3), and both sides are faced so that they are exactly parallel. The outside of the master-plate should be set perfectly true by the use of a test indicator, after which a central hole a, Fig. 2, is bored and reamed to fit a plug gage 0.050 inch in diameter. The master-plate is then

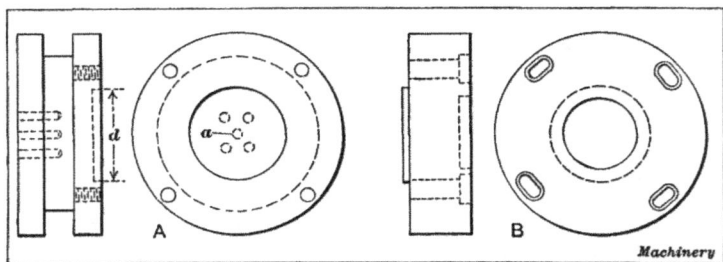

Fig. 2. Master-plate for Accurate Jig Boring

removed, and the four outer holes around the central hole a are laid out from the center hole to the dimensions given at E and F, Fig. 1. The centers of these holes are next drilled and tapped for $\frac{1}{32}$-inch screws, which are to be used for holding indicator buttons in place. Four of the buttons are fastened to the plate with small screws, and then by working from a plug in the central hole, the buttons are shifted until their center distances exactly coincide to those given. When the buttons are accurately set, the master-plate is again strapped to the faceplate, and a test indicator is used to set one of the buttons perfectly true. This one is then removed so that a hole may be bored and reamed. In the same way the other three holes are bored and reamed, and the size of each is made to coincide with that of the central hole a, Fig. 2. It will be seen that if the buttons are accurately set, the center distances of the holes will also be accurate, as they are bored concentric with the buttons.

A piece of wire *e* (Sketch *D*, Fig. 1) is next turned to fit the taper hole in a collet, as shown, after which both the collet and wire are placed in the lathe spindle, the faceplate is removed and the end of the wire is turned to a close fit for the holes in the master-plate. The faceplate is again screwed on the lathe spindle, and the master-plate is strapped to it with the wire plug *e* in its center hole. The recess *d*, Fig. 2, is now turned, after which the plate is removed so that the four $\frac{3}{16}$-inch screw holes shown

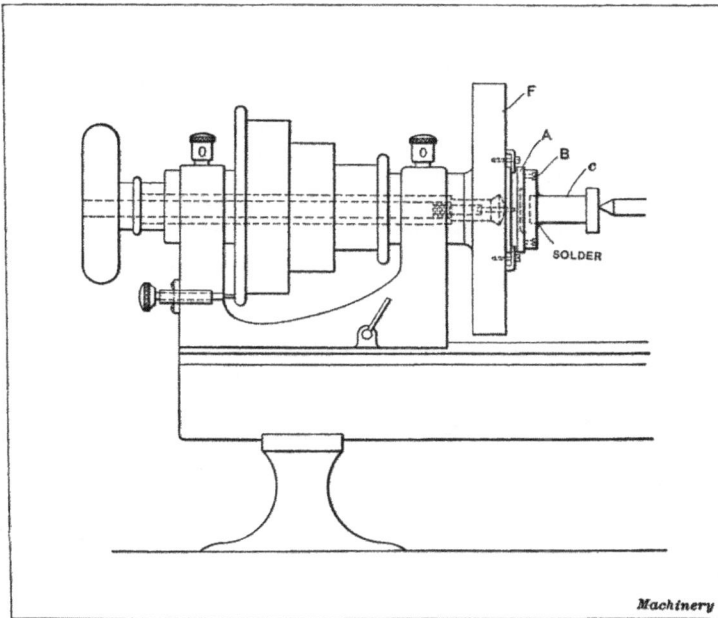

Fig. 3. Master-plate and Jig strapped to Faceplate of Bench Lathe

may be drilled and tapped in its outer flange. A brass piece *B*, Fig. 2, is next turned and a shoulder is formed on it which is an accurate fit in the recess *d* in the master-plate. Four elongated holes are also milled in plate *B* to match the holes in the flange of the master-plate. Piece *B* is then fastened to the master-plate, and the latter is strapped to the faceplate of the lathe, with the wire plug in its central hole. A recess is then bored in *B* to an accurate fit for the small end of jig *c* (Figs. 1 and 3), which is then soldered to *B*, while the tail center is against the end.

The master-plate is now shifted so that one of the four outer holes fits over the wire plug in the spindle. When the master-plate is changed to a new position, care should be taken to see that no dirt or chips get between it and the faceplate, as these would, of course, impair the accuracy of the work. One of the four holes is now bored, reamed and counterbored in jig c. In the same manner the remaining three holes are finished, and obviously, their location will correspond to the holes in the mas-

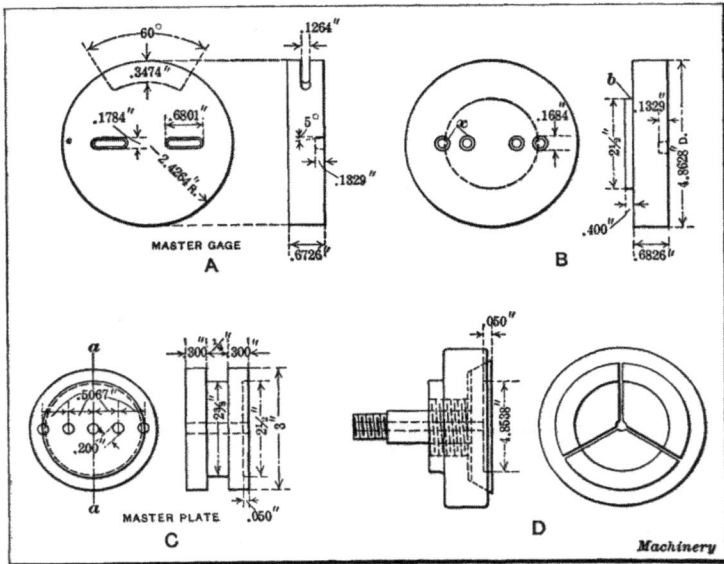

Fig. 4. Special Gage — Master-plate — Step Chuck and Closer

ter-plate. When the holes are finished, the jig is easily removed by turning off the solder with a side tool. The master-plate should now be placed in its central position and a larger recess turned in B to fit the shoulder on jig d, Fig. 1. This jig is then soldered to B, and four holes are drilled in it in the same manner as previously described.

Accurate Gage Work in the Bench Lathe. — It was not very long ago that the making of a gage such as is illustrated at A in Fig. 4, by methods which would enable practically the exact duplication of any number, was not only very expensive but quite impracticable; but now with the improved tools and methods of the modern shop, skilled workmen can produce exceedingly accurate work, but not without care and perseverance. Three of these gages were required, one for manufacturing purposes, one for the inspector, and one as a master-gage. By the simplified method to be described, each of these gages can be finished ready for hardening with one setting on the master-plate, thus insuring great accuracy. To work to the highest degree of accuracy, each operation must be positively gaged so that it is certain when the work is finished that every step has been performed correctly.

Making the Master-plate. — In order that the three gages be exact duplicates, it is first necessary to make a master-plate. The finished master-plate is shown at C in Fig. 4. This plate is made from tool steel, but is not hardened when finished. It can be rough turned in an engine lathe to within 0.020 inch of the required size and afterwards be finished in a bench lathe. When the master-plate is ready for the bench lathe, the faceplate of the latter should be tested for accuracy by the use of a test indicator. After the master-plate is strapped to the lathe faceplate, one side should be faced off with a side tool; the work is then reversed and the opposite face and outer half are machined. Both sides will then be perfectly parallel. The master-plate is now ready for the buttons which are used for locating it when boring the five holes shown.

On a center line at right angles to the line a–a, the holes for the screws which are to hold the buttons in place are laid off to the required dimensions and drilled for a $\frac{1}{8}$-inch tap. The depth of these holes should be about $\frac{1}{4}$ inch. When the five holes are tapped so that the button retaining screws may be inserted in them, the center button is screwed in place and located in the center of the plate by using a depth gage. Assuming that all the buttons are hardened, ground and lapped on the outside to the same diameter, they may be set the correct distance apart by using as a gage a flat piece of steel having a thickness equal to the center-to-center distance minus the diameter of one button. The five buttons may be aligned by laying a straightedge across them.

The master-plate is now ready to be bored and reamed. After it is strapped to the faceplate, each of the holes is bored in its correct position by the usual method of setting first one and then the other of the buttons true and then removing the button and boring the hole. The four outer holes should be bored first, the one in the center being machined last. Great care should be taken before boring a hole to set the button perfectly true. This is very important as the accuracy of the gages depends altogether on the accuracy of the master-plate. After a button has been set approximately true by the test indicator, balance weights should be fastened to the faceplate to counterbalance the work which is, of course, offset with relation to the faceplate when boring the outer holes. This is essential because if the faceplate is not perfectly balanced, it will impair the accuracy of the hole being bored. When testing the balance, the driving belt should be removed from the cone pulley.

After a button is indicated, it is removed and a center is made by the use of a V-pointed spotting tool. A No. 15 drill, in this case, is fed through the master-plate after which the hole is bored out to 0.198 inch with a boring tool, which is followed by a 0.200-inch reamer, thus finishing the hole to the size indicated at C, Fig. 4. In the same manner the other four holes are spotted, bored and reamed to the same size. After the central hole is finished, a recess 2.5 inches in diameter and 0.050 inch deep should be bored at the same setting of the work. A light cut should also be taken on the outer diameter with a diamond-point tool.

First Operation on Gage. — After the gages are roughed out in an engine lathe to within 0.020 inch, they should be annealed by being packed in charcoal in a cast-iron box with a close-fitting cover luted with fireclay in order to exclude the air so that the gages will not change too much when they are hardened. The iron box should be placed in the furnace and given a very slow heat for about three hours, after which the heat is shut off and the box is left in the furnace over night. On removing the gages, they will be found very soft and easy to work.

The gages are now ready for the first operation in the bench lathe. First strap one of them to the faceplate with the boss b

out (see sketch *B*, Fig. 4) and face this end. Then machine the boss *b* to an accurate turning fit in the recess previously bored in the master-plate. The gage should then be removed and clamped to the master-plate with two C-clamps, so that the two may be soldered together. Before soldering all oil should be removed from the surfaces, which may be done by spraying a little benzine over them and afterward wiping it off with a piece of clean waste. When soldering, the gage should be heated over a Bunsen burner until the solder will run freely. Plenty of solder should be used at the point indicated in Fig. 5. When the parts are cooled they will be securely attached to each other and the C-clamps can be removed.

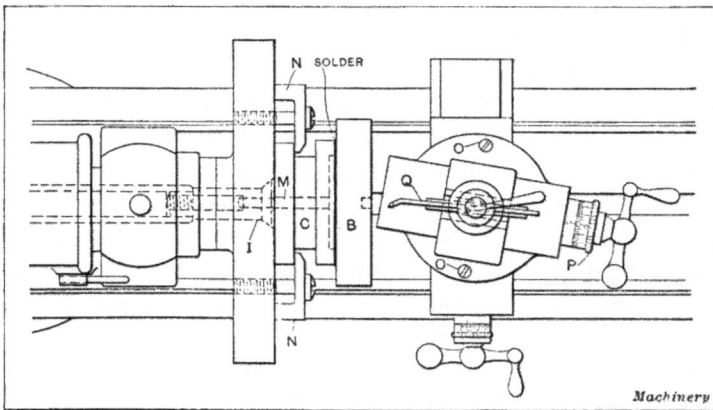

Fig. 5. Plan View of the Bench Lathe arranged for Boring the Tapered Holes in Face of Special Gage

The plug collet *I* is next placed in the lathe spindle and the handle is tightened. A tool steel plug *M*, tapered at one end, is then driven lightly into the collet. This plug is turned to 0.2003 inch in diameter or 0.0003 inch larger than the master-plate holes. Care should be taken to see that the shoulder on pin *M* is below the surface of the faceplate. The pin *M* is next polished with fine emery cloth until it is a turning fit in the holes in the master-plate. The latter is then fastened to the faceplate with four straps *N*, the pin *M* being inserted in the central hole. If the spindle does not balance perfectly, a small weight should be attached to it. The gage is now faced to a thickness of 0.6826 inch and turned to a diameter of 4.8628 inch. The work is now removed and the faceplate cleaned thoroughly, after which the master-plate is again attached to the faceplate with the center plug M in any one of the four outer holes. The spotting tool is then used to cut a center for starting a No. 25 drill, which is sunk to a depth of 0.122 inch. The compound rest is then swung to an angle of 5 degrees, as shown in Fig. 5, by the loosening of the screws O. When the point of the boring tool Q is in contact with the surface of the gage, the graduated sleeve P is set to zero and a taper hole is bored to a depth of 0.132 inch and to a diameter on the outside of 0.1684 inch. These dimensions can be obtained very accurately by using a tapered plug-gage having a shoulder which is allowed to clear the work about 0.005 inch for grinding. In the same manner, the three remaining holes which are to form the ends of the slots shown in the face of the gage at A, Fig. 4, are bored. Care should be taken to balance the spindle each time the master-plate is shifted from one hole to another.

Milling Taper Grooves in Bench Lathe. — After the four tapered holes U and V (see Fig. 6) have been bored, as described in the foregoing, the bridges b are milled away to form slots or grooves with tapering sides. The indexing attachment is first placed on the lathe headstock. The indexing plate C has 360 divisions. This is inserted by first removing the handle K. In order to make room for the block E with its indexing pawl D, the headstock is moved along the bed by loosening handle A. Tapered plugs having outer ends which are parallel and of the same size should be placed in the two extreme holes in the gage. The master-plate (with gage still soldered to it) is then clamped to the faceplate with the plug M in the central hole. The pawl D should be inserted in the zero groove of the index-plate. An angle iron Z is next fastened in the T-groove on the back of the lathe bed. On top of this angle iron the milling attachment J is placed and adjusted so that the spindle will reach the two outer holes in the gage when it is moved to and fro by lever B, which is attached to stud O and fulcrumed on the stud of extension block P, which is fastened to the bed. To permit this movement of the

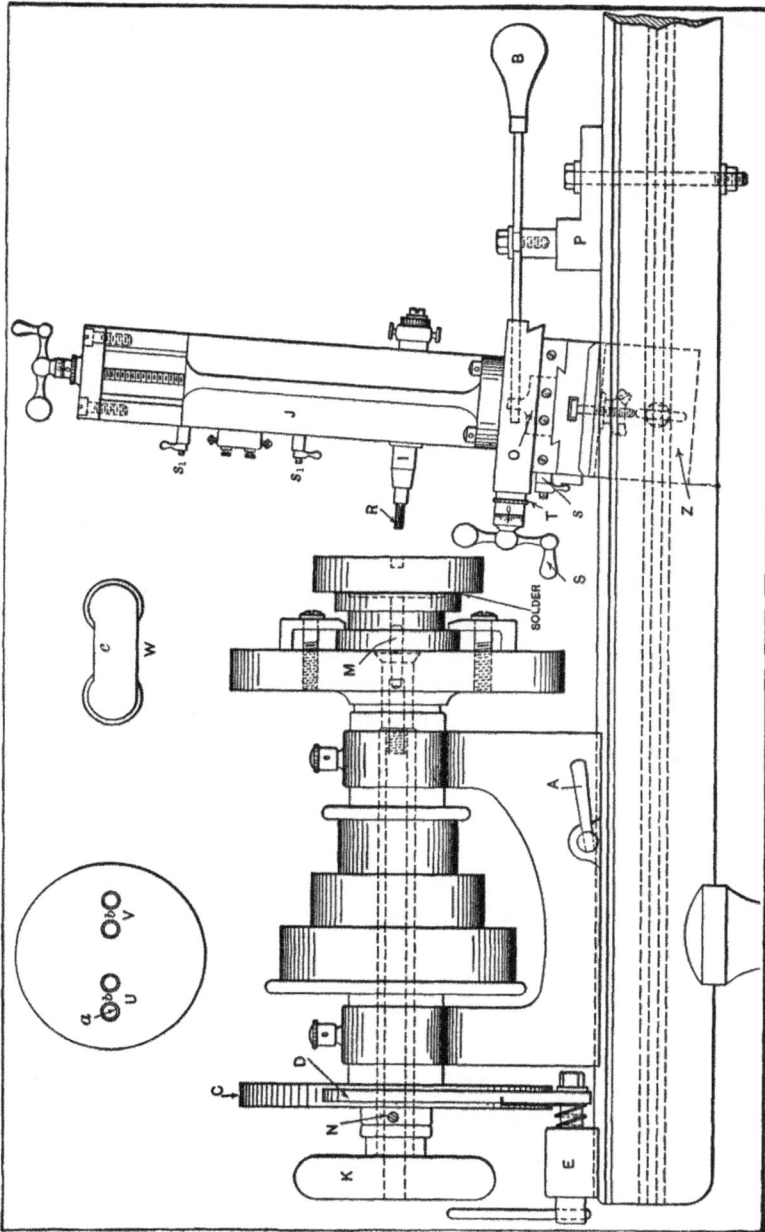

Fig. 6. Bench Lathe equipped with Indexing and Milling Attachments

slide by lever B, the cross-feed screw is removed. The plugs previously inserted in the end holes of the gage are now used to set the holes in a horizontal position or parallel with the movement of the cross-slide. This is done by inserting a test indicator in the spindle I of the attachment and moving the indicator across the plugs by lever B. When the indicator shows that each plug is exactly the same height, the screw N of the index-plate C should be tightened. The milling attachment should also be adjusted until the indicator shows that it is parallel with the face of the gage.

The end mill R, 0.150 inch in diameter, is now inserted in the spindle and the movable stops s on each side of a stationary stop on the horizontal slide, are set so that the mill has a horizontal movement, approximately equal to the center-to-center distance between the holes U. The mill is also set central vertically with these holes. The bridge b is then cut away by feeding the mill to and fro with lever B and inward by handle S, thus taking a number of successive cuts. When the end of the mill is brought in contact with bridge b, before taking the first cut, the graduated sleeve T should be set to zero so that it may be used to determine the depth of the slot. After the bridge b has been cut away, the mill is fed downward until it just comes in contact with the bottom corner a of one of the holes. The movable stops s_1 on the vertical slide are now used to lock the mill against vertical movement. The mill is then fed across as before, machining the bottom of one slot as shown in the enlarged detail at W. The work is indexed 180 degrees and the milling operation described in the foregoing is repeated on the other pair of holes. Each slot would then appear as at W. In order that the sides c be cut away, the end mill is shifted so that it operates on the opposite side of the lathe center. The stops s on the horizontal slide are readjusted for this new position, the vertical stops remaining the same as before. After one side c has been milled the work is again indexed 180 degrees and the milling operation is repeated.

The slots will now have tapering ends but straight sides. To mill the sides to a taper corresponding with that of the ends, the milling fixture is inclined to an angle of 5 degrees, as shown in

Fig. 6. The end mill should then be replaced with an indicator, which is again used to test the parallelism of the face of the gage with the milling attachment. The sides of the slots can now be milled tapering in precisely the same manner as described for the end milling operation; that is, by milling one side, then indexing the work 180 degrees, and finally finishing the two remaining sides with the mill on the opposite side of the center. Very light cuts must be taken during this operation.

Milling a Circular Groove in Bench Lathe. — Before milling the circular groove in the periphery of the gage shown at *A*, Fig. 4, a dividing head should be attached to the lathe as shown in Fig. 7. The index-plate *C*, Fig. 6, is replaced with a worm-wheel *A*. The worm meshing with this wheel should be set central by adjusting the bracket *C* along the lathe bed. An extension block *I* is also inserted beneath the milling attachment to raise it to the required height. The milling fixture is now used as a vertical miller by changing the position of the spindle as shown. An indicator is next placed in the spindle *G*, and with indexing handle *E* in the zero hole of the index-plate, the two slots which were previously milled in the face of the gage, are set parallel with the movement of the cross-slide by turning the lathe spindle; the screw *B* in the worm-wheel is then tightened. A circular mill of the required size is next inserted in the spindle. This mill must be set central with the sides and periphery of the gage. This can be done in the following manner: first lower the mill, which should be revolving; then turn handle *P* until a "listener" indicates that the cutter is barely in contact with the side of the work. Graduated sleeve *Q* should then be set to zero, after which the mill is raised and moved inward a distance equal to one-half the width of the gage plus one-half the diameter of the end mill. The latter is now central with the sides of the gage. By the same method, it may also be set central with the periphery. The work is now indexed 30 degrees, which will require with this particular dividing head, 15 turns of the indexing crank. A stop *F*, one of which is attached to the flange of the cone pulley on each side of the headstock, is then set against its striking point on the headstock. The work is then indexed 60 degrees or 30 turns in the opposite

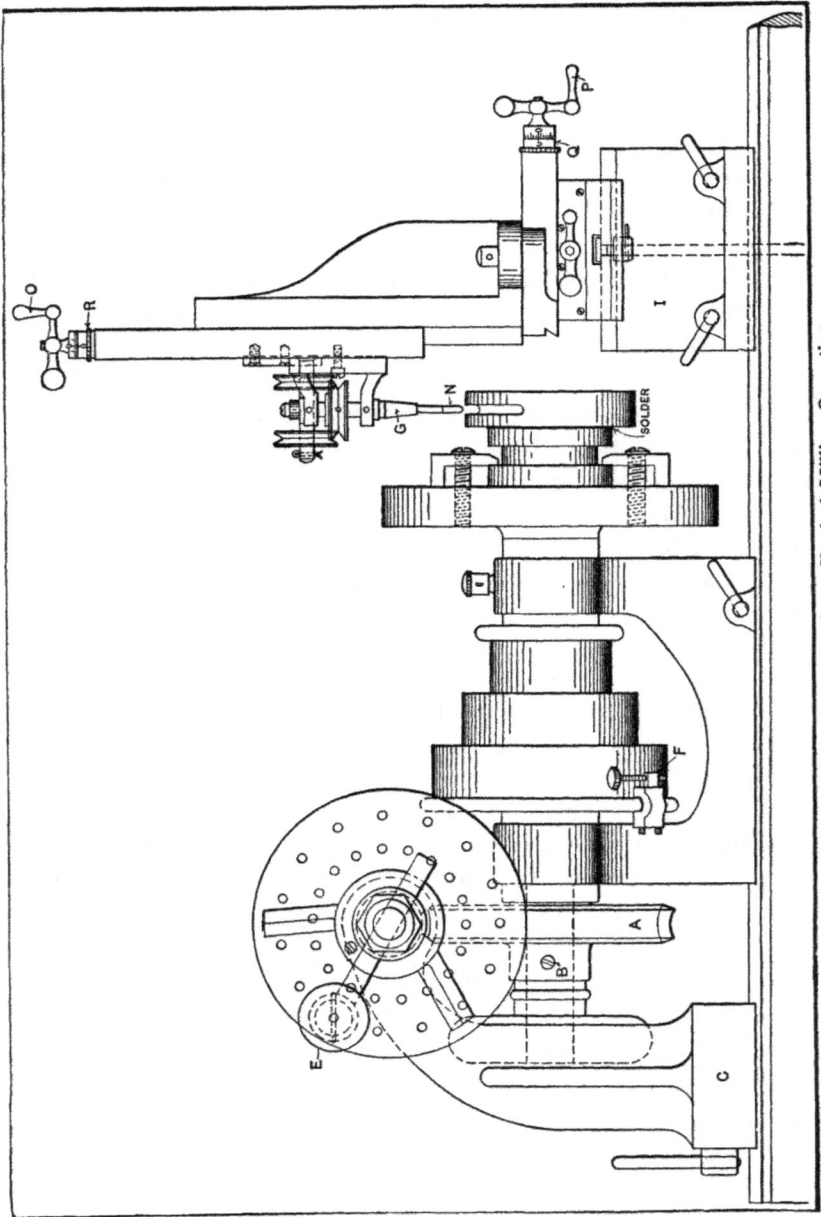

Fig. 7. Bench Lathe set up for a Vertical Milling Operation

direction, after which the stop F on the opposite side is adjusted as before. These stops, in each case, should be so set that the indexing pin will just enter the zero hole of the dividing plate.

Everything is now ready for milling the circular groove which, as shown in Fig. 4, has a length of 60 degrees. Prior to the milling operation, however, a hole 0.335 inch deep, as indicated by the sleeve graduations, should be drilled at each end of the slot with a No. 35 drill. These holes will provide a clearance space for the end mill. A square end mill of 0.110 inch in diameter is then used to rough out the groove to a depth of about 0.295 inch. The circular end mill is then employed for finishing the groove to the required depth and width. The work is fed by turning the indexing crank, and the stops F act as a positive gage, thus insuring that the slot will be the required length. The milling operation is now completed. After the attachments have been removed and the lathe arranged for turning, the gage should be removed from the master-plate by cutting away the solder with a side tool. One gage is now complete and ready for hardening. By the use of this same master-plate, obviously, the other two gages may be made precisely in the same manner as described in the foregoing.

Grinding Special Gages in Bench Lathe. — The grinding operations on the gage illustrated at A, Fig. 4, are performed in practically the same way as the milling or turning, the only difference being that a wheel is used instead of a tool or cutter. The gage is, of course, mounted on the master-plate as before. Great care must be exercised before soldering the gage to the plate, that it be located in the proper position relative to the holes in the master-plate. This may be done as follows: First clamp the master-plate and gage together, with two C-clamps, then place a close-fitting plug in each of the two extreme holes in the master-plate. These plugs, which should extend about one inch beyond the surface of the plate, are next placed on the top surface of a parallel block which should be high enough to clear the diameter of the gage. This block should be mounted on an accurate surface plate so that a test indicator may be used to set both grooves in the face of the gage parallel with the plugs in the master-plate. After the grooves have been tested on one side, the master-plate should be turned over, so that the opposite sides of the plugs rest on the parallel block. If the indicator then shows the same reading as before, the gage may be soldered to the plate. The latter is then mounted on the faceplate of the lathe with the plug M (Fig. 5) in the central hole, and the gage is indicated on its face and periphery. If it runs out more than 0.004 inch it should be reset. If, however, the error is small enough to permit truing, rough grind the face and periphery leaving about 0.001 or 0.002 inch for finishing after the grooves are ground. This precaution must be taken to insure accuracy, as hardened steel changes during a grinding operation after the outer scale is removed.

In grinding the ends of the grooves, a diamond lap is used. The end of this lap is charged as well as the body. When the ends are being ground, the slide-rest is set to an angle of 5 degrees and the procedure is practically the same as when the ends of the slots were bored; 0.0002 inch should be left for hand lapping. The diamond lap should run at the fastest speed obtainable and the work should revolve at the slowest. To grind the sides of the grooves, the same diamond lap is used and the machine is set up as shown in Fig. 6. A circular end diamond lap 0.115 inch in diameter is used for grinding the circular groove, and the machine is equipped with the same attachments as are illustrated in Fig. 7.

After the periphery and one side of the gage have been finished, the side next to the master-plate still remains to be ground. A step chuck and closer D (Fig. 4) is placed in the headstock spindle and three small pieces of sheet steel of the same thickness as the slots in the chuck are placed in each slot. The chuck is then tightened, after which a recess about 0.050 inch deep and large enough in diameter to be a snug fit for the gage, is bored. The chuck is then loosened and the sheet steel pieces removed. The ground face of the work is next inserted in this recess, which should be very carefully cleaned, and the chuck is tightened. Care should also be taken to have the inner face of the gage against the chuck. If the work when tested on the periphery runs true, it may then be finished, first by using a coarse carborundum wheel and afterwards a No. 120 emery wheel; 0.0002 inch should again be allowed for lapping.

When the three gages have been ground as described in the foregoing, the faces should be lapped by rubbing them on an accurate surface plate, using a very little fine emery with plenty of benzine. A smooth and bright finish may be obtained by using a dry and clean plate. In this way, each gage is brought to the exact measurements required. The grooves are draw-

Fig. 8. Dividing Head and Tool-rest used in Laying out and Boring Master-plates

lapped with a flat copper stick, powdered emery and sperm oil being used. For a final finish on very accurate work, the copper lap should be followed by one made of boxwood, the abrasive being white powdered oilstone.

Boring Master-plates in Bench Lathe. — For laying out, drilling and boring master-plates, the special dividing head and swing tool carriage shown in Fig. 8 are used in the factory of the Illinois Watch Co. The dividing head A consists of a false faceplate B, which is fastened to a cross-slide and which, in turn, is attached at right angles to another cross-slide, the guides of which are solid with the faceplate proper. The whole mechanism is mounted in a bearing of its own and is driven by means of the dog E, which engages a slot in the lathe faceplate, as shown.   The plate B is graduated on its periphery in degrees, and it can easily be turned on its center and clamped at any desired angle.   The cross-slides are both fitted with micrometer screws which afford quick and accurate adjustments of the plate in either direction along their line of travel.   As can easily be seen the tool carriage F has a part G which is hinged at H and which may be adjusted out or in by means of the micrometer head K, mounted in part I, which is solid with the base of the

Fig. 9.   Bench Lathe equipped with Quill used when Boring Center Holes in Watch Plates

carriage.   The center of the tool-holder or spindle M in which the boring tool is placed, is exactly half way between the center of the pin H and the center of the micrometer head K, so that if the micrometer screw is turned outward 0.001 inch the tool is set out just half as much, or 0.0005; in other words, the direct reading of the micrometer indicates the amount that the bored hole will be enlarged on the next cut.   The tool is fed into and out of the work by turning the handle P.

Formerly all master-plates were originally laid out with reference to two center-lines at right angles to each other, the location of each hole being indicated by the horizontal and vertical dimensions from the center-lines. A method which is considered much simpler and less liable to cause error is now used for this class of work. When laying out the master-plate, all holes that are equi-distant from the center are connected by arcs passing through their centers; the radius of each arc is then given. The position of the holes on the arcs is given in degrees and minutes from a center-line marked zero. The advantage of this last method to the toolmaker will be apparent to anyone who has ever worked out a complicated plate by the first method referred to.

Fig. 10. Milling Attachment for Bench Lathe used for Milling Small Punches, Dies, etc.

The mounting of the dividing head of the machine shown in Fig. 8, in its own bearing, illustrates the general principle upon which all fixtures used in the factory to hold parts to be bored are made. The usual form of individual bearing, or quill, as it is called, when used with a separate holder, is shown in Fig. 9. The quill illustrated in this engraving is used to hold a watch-plate while the center hole is bored. The quill holder is in the form of a V-block and clamp, and the quill is driven from the lathe spindle by a tongue and fork, as shown at A. Each train hole in a watch-

plate has a separate quill for centering it while it is being bored
out, the plate being located in the jig which holds it, by means of
the dial feet holes.   There are pins in the face of each quill so
located that when they enter the previously drilled dial feet holes,
the plate to be drilled is brought into the proper position.   The
watch-plates are all first spotted for the drills, then drilled and
finally bored.

Milling, Filing, and Grinding Attachments. — Fig. 10 shows
a milling attachment used on a bench lathe for working out slots
and irregular places in small jigs, punches and the like.   When

Fig. 11.   Vertical Milling Attachment for Machining Die Stripper
Plates, etc.

using this milling attachment the lathe spindle is locked and the
spindle of the attachment is driven by a round belt passing over
pulley A.   Several very interesting attachments for the bench
lathe are used in the punch and die department of the Illinois
Watch Co., for handling the special work.   Fig. 11 shows a table
with a small end mill A in the center, which is used for some classes
of die work.   The milling cutter is rotated by means of a pair of
bevel gears and a shaft connected to the lathe spindle.   Fig. 12
shows a filing machine, the table of which can be tilted to a limited

extent for the purpose of filing clearance in a hole or for other reasons. The stroke of the file is adjusted by shifting the crank-pin in the slotted plate which is attached to the nose of the lathe spindle. Fig. 13 shows an adjustable plate used when surfacing off small punches or dies with an emery wheel; this plate is extremely handy for many other light grinding jobs.

Setting Slide-rest for Cylindrical Grinding. — To set the slide-rest of a bench lathe to grind parallel or cylindrical, a cylindrical test piece having centers in each end and a collar on one end may be used and accurate results obtained. The collar is first ground

Fig. 12. Bench Lathe Filing Attachment for Filing Clearance in Dies, etc.

true, after which the position of the test piece is reversed. When the emery wheel is moved up to the other end, its position with relation to the collar as indicated by the sparks, will make it possible to set the slide-rest very accurately, though it may be necessary to repeat this cut-and-try method several times.

When finishing holes by grinding, particularly when the work is hardened and is to be lapped, it is especially desirable to get the hole as nearly parallel as possible. This may be done by feeding the emery wheel through the hole and making it cut first on one side and then on the other.   By noting the sparking of the wheel, the slide can be set so that the hole can be ground parallel, to an extreme degree of accuracy.   If the hole is very deep or small in diameter so that the sparks cannot be seen, a "listener" is used which consists of a piece of small drill rod with a handle on one end.   The end of the rod is applied to some part of the grinder and the handle is held close to the ear.   By the sound emitted when the wheel is in contact with the work, it is easy to tell if the cut is the same throughout the length of the hole on both sides. If this "listener" is not sensitive enough, the bottom of a tin can

Fig. 13.   Adjustable Plate for Surface Grinding on the Bench Lathe

(having one end open) may be applied to some part of the grinder or to the top of the tool-post screw.   The contact of the emery wheel with the work can then be heard easily by listening at the open end of the can, even though the wheel is not taking a cut deep enough to make a visible mark in the hole.   These methods are not only better, but far quicker than calipering for parallelism.

Internal Grinding in Bench Lathe. — The making of the different sizes of draw-back chucks for the bench lathes manufactured by Hardinge Bros. requires an accurate internal grinding operation.   Chucks ranging in size from 0.010 inch up to $\frac{3}{16}$

inch are finished by lapping, while the larger sizes are ground with the special wheels. In finishing these chucks, they are mounted in the lathe spindle in the regular way, and the grinding attachment drives the wheel. Before hardening, the chucks are bored to approximately the required size, and, in the cases of the smaller sizes, which are to be finished by lapping, the holes are made to within about 0.001 inch of the standard sizes.

The plugs used for the lapping operation are made of soft steel and are driven at speeds ranging from 4000 to 5000 revolu-

Fig. 14. Lapping a Small Chuck or Collet in Bench Lathe

tions per minute. Fig. 14 shows a machine equipped for the lapping operation on a chuck of 0.125 inch internal diameter. The lap is driven at 4500 revolutions per minute and is moved in and out of the hole by means of the system of levers connecting the spindle with the rod at the right-hand side of the machine, which rod is given a reciprocating movement by a crank on a shaft behind the bench. The important point in securing accurate work in these lapping operations is to have the work so heat-treated that all parts are of equal hardness. If this precaution is not observed, the softer parts of the work will take up the abrasive and thus resist being cut away; the lap will also be damaged through the "back-biting" action of the abrasive imbedded in the work.

An important advantage is secured in internal grinding operations through the use of the push spindle, which is commonly used on bench lathes for work of this kind. Fig. 15 shows a "Cataract" precision bench lathe equipped with one of these push

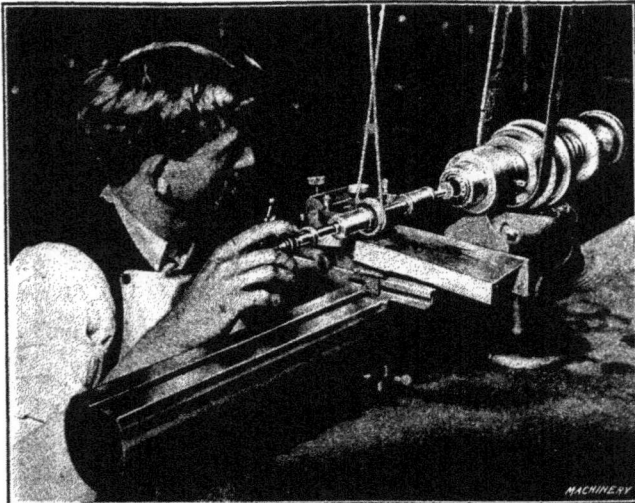

Fig. 15.   Method of Using the Push Spindle Grinding Attachment

spindle grinding attachments. This spindle is held by two bearings provided with dust-proof washers to prevent particles of abrasive material from damaging them, and may be advanced to the work by pushing forward the finger piece, as the illustration indicates. This finger piece is held between the thumb and index finger, as shown, and the sensitive touch which the operator secures in this way enables him to tell whether he is working under proper conditions, as soon as the wheel comes into contact with the work. The sparks thrown by the wheel also act as a guide in enabling the operator to tell whether the wheel is cutting properly. Greater precision is made possible through the use of the push spindle, and much of the trouble experienced with broken wheels and spoiled work, when using the ordinary spindles that are advanced to the work by means of reciprocating slides, is avoided.

Bench Lathe Turret Attachment. — In the manufacture of small screws, studs, and many other pieces on which there is a series of operations, such as centering, drilling, and tapping, a marked advantage is secured by having a series of tools available for performing successive operations on a piece of work at one setting. In order to meet the requirements of such work, bench lathes are sometimes equipped with a turret attachment. The "Cataract" bench lathe shown in Fig. 16 has a six-hole turret attachment and a double tool cross-slide, thus giving the machine a capacity for eight tools. The turret is mounted on a slide and is provided with rack and pinion feed, as this design has been found to be best suited for the requirements of the sensitive work which is done on a turret of this size. An adjustable stop is placed at the back of the slide to check the tools at the desired point, or independent stops may be used for each tool. The double tool cross-slide has a T-slot running across it in which the two toolposts are mounted. The cross-feed of the slide is operated by means of a hand lever, as the illustration indicates, and adjustable stops are located at each end of the slide.

An automatic chuck-closer is used in connection with the turret and double tool cross-slide when operating on bar stock. This device is used in place of the regular draw-in spindle, and closes the chuck by simply throwing over a hand lever. The chuck-closer fits over the outside of the spindle at the rear end and is keyed in position. It enables the machine to be run continuously, without stopping to adjust the chuck, the stock being fed through the spindle as required.

The bench lathe shown in Fig. 16 is equipped for making small brass washers for clock movements. These washers are made from bar stock; they are 0.281 inch in diameter, 0.057 inch wide, and have a central hole 0.156 inch diameter. The washer is beveled to an angle of 45 degrees on one side. Three tools are used in the turret, namely, a stop, a centering tool, and a drill for the hole. A forming tool for beveling the edge of the washer, and a parting tool for cutting it off from the bar, are used in the double tool cross-slide.

With the stock in the spindle and the attachments adjusted, the operator throws over the lever of the turret attachment with his right-hand. This brings the stop against the end of the stock and pushes it back through the spindle until the stop on the turret slide is reached. The operator then closes the chuck by throwing over the lever on the automatic chuck-closer with his

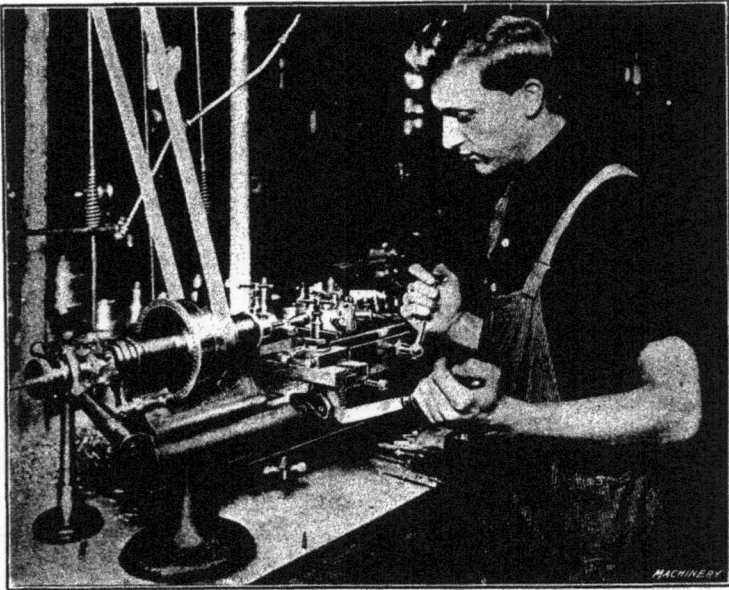

Fig. 16. Bench Lathe equipped with Turret Attachment, Double Tool Cross-slide and Chuck Closer

left-hand. He then centers the stock and drills the hole. The lever of the double tool cross-slide is then pushed down to bring the forming tool into contact with the work to bevel the washer. The final step consists in pulling up the lever on the cross-slide to bring the parting tool into action to cut off the finished washer from the bar.

It will be noticed from the preceding description that the stops on the attachments regulate all of the operations, so that the operator does not have to pay any attention to the dimensions of the work. This greatly increases the output on work of this character. On the operation just described, an average workman can turn out three hundred washers per hour. It is a relatively easy matter to design the necessary tools to handle any given piece of work, and after such tools have been made, a turret lathe equipment of this kind can be used to great advantage in the manufacture of many pieces that are now being produced in a far slower way.

Thread Chasing Attachment. — The thread chasing attachment which has been designed by Hardinge Bros. for use with

Fig. 17. Bench Lathe equipped with Thread Chasing Attachment —
Rear View

the bench lathe, is shown applied to a lathe in Fig. 17. It is bolted to the faced and T-slotted surface at the back of the bed, adjustable brackets and eccentric clamp bolts being used for this purpose. The chasing bar extends the entire length of the bed and the tool-holder is adjustable along the chasing bar so that it can operate in any position over the bed. This tool-holder carries circular cutters for both inside and outside thread chasing, and has a finely graduated feed-screw for making adjustments. The regular cutters used with this attachment work in a hole of from $\frac{1}{2}$ inch up, but special cutters may be made to chase a thread on the inside of a much smaller hole. A stop-collar on the chasing bar may be set so that it stops the cutter at any desired point, and this makes it possible to cut a clean thread right up to a shoulder on either inside or outside work.

The lead-screw used with this attachment is fluted for a distance of 1 inch from the end, thus forming a hob for cutting the leading nut. In cutting a nut in this way the blank is fastened

Fig. 18. Cutting a Special Tap in Bench Lathe with the Screw-cutting Attachment

in the socket carried by the chasing bar, and is then held against the hob, which is driven by the regular gearing provided for driving the attachment. The lathe can be geared to cut from one to eight integral multiples of the number of threads on the lead-screw.

This attachment provides a rapid method of cutting threads, as the leading nut can be lifted out of engagement with the screw to allow the tool to be quickly returned by hand to the beginning of the cut, without the necessity of either stopping or reversing the lathe. The attachment is so arranged that it does not interfere with the regular lathe work or with that of any other attachment; consequently it can always be kept mounted on the machine, where it is ready for immediate use.

Screw Cutting Attachment. — A screw cutting attachment is illustrated in Fig. 18, engaged in making a special form of tap. It will be noted that this particular attachment consists of a universal-jointed rod, which is geared to the lathe spindle by gears mounted on a bracket screwed to the end of the bed. Change gears are provided with this attachment which make it possible to cut from 5 to 150 threads per inch. The feed of the tool over the work is obtained by means of the universal-jointed rod, which fits over a nut on the end of the longitudinal feed-screw in the regular compound rest, and as this compound rest may be set to any desired angle by the graduated swivel, it is possible to cut a thread on any taper.

Thread Milling Attachment. — The bench lathe equipped with a thread milling attachment has been found especially useful for milling small worms, threaded parts of meters, and for a variety of other similar work in the Hardinge Bros. shop. This attachment is illustrated in Fig. 19, engaged in threading a part of a meter mechanism. The lathe equipped for screw cutting, as illustrated in Fig. 18, provides for the feed of the milling cutter, which is carried on a spindle running in a bearing bolted to the top of the compound rest. The driving pulley on this spindle is belted to a pulley on the countershaft, as indicated in the illustration. The speed of the work is regulated by the worm-wheel mounted on the lathe spindle nose. This worm-wheel is driven by a worm which, in turn, is driven by the pair of bevel gears and a cone pulley belted to the countershaft, as shown. The bracket which carries the bevel gears and cone pulley is held in the same T-slot that carries the thread chasing attachment when the latter is in use.

An automatic stop is provided which can be set to throw the worm out of engagement with the wheel and thus stop the feed of the cutter at any desired point. A little ingenuity on the part of the toolmaker will enable him to vary the minor details of this equipment so that he can accomplish anything in thread milling within the capacity of the machine.

Use of Bench Lathe for Manufacturing. — The bench lathe is generally regarded as a machine tool adapted only for light turning, boring, milling and grinding, and one particularly suited to work required to be finished within close limits — hence the common designation of "precision" lathe. Of course, this generalization is not strictly true, inasmuch as there are engine

Fig. 19. Thread Milling Attachment applied to a Bench Lathe

lathes made with short legs and pan for use on benches. These are lower-priced than regular bench lathes of the precision class and are not regarded as being so well suited for finishing small parts with speed and accuracy as the special type of lathes. The fact that the so-called "precision lathe" is a machine tool of wide adaptability which can be made highly efficient for certain classes of manufacturing as well as for a wide range of tool-room work,

is not as well appreciated as it should be.   The factors in favor
of the bench lathe are compactness, light weight of moving parts
and the possibility of running the spindle at high speeds continuously without serious heating.   The high speeds with the cuts
and feeds possible with the best high-speed steel tools enable an
expert operator to show a chip production in a day's work that
compares favorably with the average production of a fourteen-
inch or sixteen-inch engine lathe working on much heavier pieces.

The modern precision lathe can be employed efficiently in the
tool-room for such work as making jig bushings.   This is a job

Fig. 20.   Precision Lathe Grinding Hardened Steel Washers

requiring accurate turning and boring, grinding after hardening,
and lapping as a finishing operation.   Other jobs such as making
small hardened and ground steel washers can be handled on a
bench lathe fitted with a grinding attachment.

The illustration, Fig. 20, shows a Cincinnati precision lathe
being used as a grinding machine for finishing some hardened
washers which were turned and bored on this machine.   The
washers are about $2\frac{1}{2}$ inches outside diameter, the hole diameter
being $\frac{7}{8}$ inch.   The lathe has a swivel headstock which may be

readily set for face grinding at such positions as are most convenient for the operator. The claim is made that with the spindle
construction and bevel gear drive employed, unusually smooth
work can be produced — that a surface for most practical purposes equal to a lapped surface can be ground in a fraction of the
time required for grinding and lapping with the common means
and methods.

Truing a Bench Lathe Bed. — The bed of a lathe that is much
used naturally wears hollow just in front of the headstock, and
this is particularly the case with a bench lathe, as this type is
used largely for chuck work. A method of doing this work,
which proved to be accurate, is described in the following. A
cast-iron templet of the shape shown in Fig. 21 was first made.

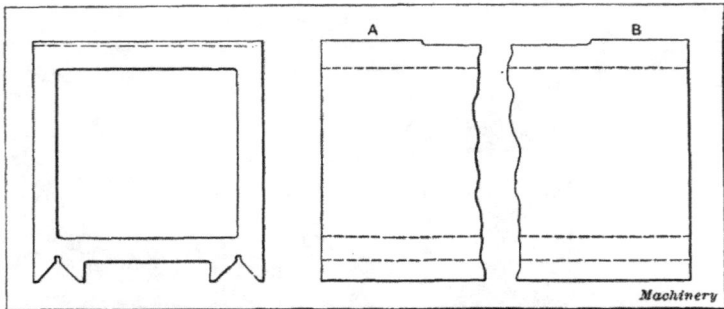

Fig. 21.   Master Surface-plate used in Scraping the Ways of Bench
Lathe Bed

The length of this templet was about two-thirds that of the lathe
bed, and two V-grooves were planed in it that were duplicates of
the grooves in the bottom of the tailstock. The templet was
made in the form shown so that it would resist bending or torsion,
and yet not be very heavy.

The distance between the V's of the lathe bed was determined
by means of a microscope held vertically in the slide-rest as shown
in Fig. 22. A brass block *A* was made with a V-groove in the
bottom to fit the V's on the lathe bed. On the top of this block
two fine lines were scribed, one being parallel to the lathe bed,
and the other at right angles to the first, but not quite touching
it. This second line was to insure the observations both being

taken at the same part of the first line, in order to eliminate any
error that might arise from want of parallelism between this line
and the bed.   The block *A* was first laid on the front V and the
microscope sighted on the marks on its surface, which appeared
through the microscope as shown at *B*.   As indicated, the eye-
piece of the microscope has cross lines arranged like the letter X.
By adjusting the cross feed-screw and the block *A* along the bed,
the intersection of these lines was made to coincide with one
edge of the longitudinal line at a point opposite the transverse

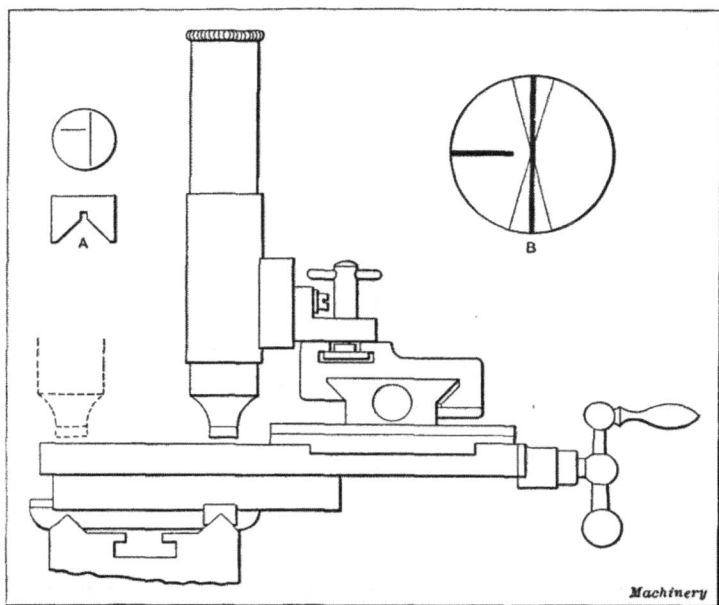

**Fig. 22.   Obtaining the Distance between the Ways of a Bench
Lathe Bed by the Use of a Microscope**

line.   When the block and microscope were set, the reading of
the cross-feed micrometer was noted.   The block was then placed
on the back V of the lathe bed, after which the cross-slide was
moved until the intersection of the cross lines again coincided
with the line on the block as before.   The reading of the cross-
feed micrometer was again taken, and a comparison of the two
readings noted, allowance being made for the number of complete
turns that had taken place.   By this method the distance be-

tween the V's from apex to apex was obtained. A gage with hardened and lapped ends was then made, of $\frac{5}{16}$-inch drill rod, to exactly this dimension. This gage was enclosed in a wooden jacket as shown in the end view at A, Fig. 23, to prevent changes in its length from the heat due to handling. The lower side of the jacket, and also the side against which the strip a is screwed, were both made parallel to the length of the drill rod, so that when the enclosed gage was laid on the cross-rail it would be parallel with the latter.

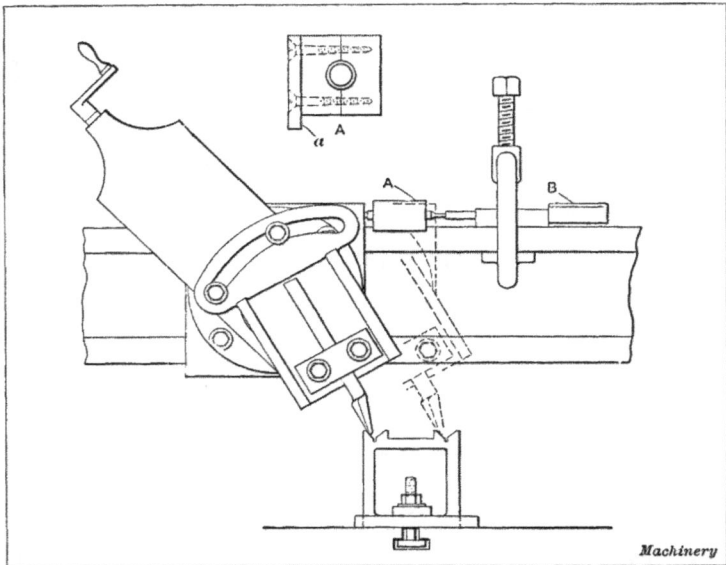

Fig. 23.   Planing the V's in the Master Surface-plate or Templet

The templet casting was first placed on the planer in the position shown in Fig. 21, and the feet A and B planed. Care was taken not to clamp the casting down any tighter than was necessary during this operation. It was then turned over so that the planed surfaces rested on the platen to which red lead had been applied in order to determine if both feet touched it throughout their entire area. This was found to be the case, so the V's of the templet were next planed. The dial indicator B was clamped to the cross-rail of the planer as shown in Fig. 23, and the saddle brought into contact with it while planing one V. The gage *A* was then interposed between the saddle and the indicator while planing the other V, the saddle being so adjusted for the finishing cuts that the indicator read the same in both cases. When the head was swung over to plane the other slope of the V's, of course the indicator had to be moved to a new position on the cross-rail. The same scheme was also used to space the two positions of the

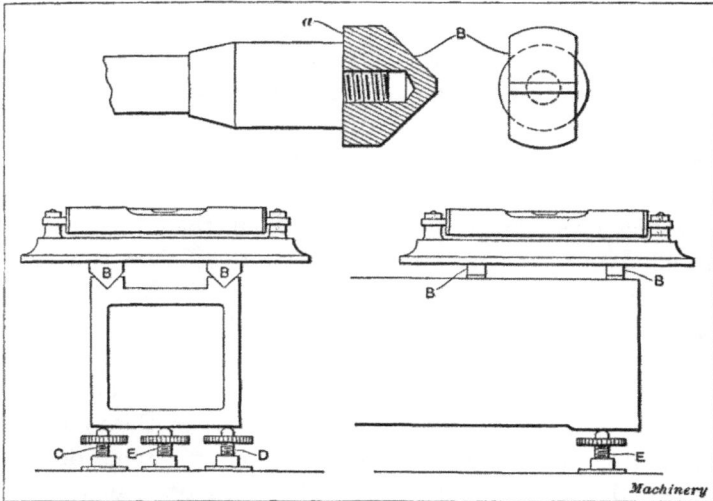

Fig. 24. Testing Accuracy of Master Surface-plate with Level

parting tool that was used to cut out the bottom of the grooves, preliminary to planing the V's.

After finishing the planer work on the templet, it was tested with a spirit level both for wind and straightness. Two pieces of $\frac{3}{8}$- by $\frac{5}{8}$-inch rolled brass were drilled, tapped, and the face *a* (see upper view, Fig. 24) carefully turned at one chucking; these pieces were then screwed on the chuck. The finishing cuts for both pieces were taken with the same setting of the milling cutter so as to make them exact duplicates in regard to the location of the V-surfaces in relation to the face *a*. The method of testing the templet for wind is shown by the lower views, Fig. 24. The templet was supported on the bench by three leveling screws, *C* and *D* being at one end while *E* was at the other. The two brass pieces B were then laid in the V-grooves near one end of the casting, and a ground spirit level laid on top of them as shown by the left-hand view. The screws C and D were then adjusted until one end of the bubble coincided with one of the marks on the vial. Then the blocks B were moved to the other end of the casting, and the level applied as before. The bubble moved to the mark with which it coincided before, thus showing that there was no wind in the V's that could be detected with that particular level. It was not a precision level, but was a great deal better than the ordinary blown level. The vial was ground true on the inside, while the outside was marked transversely by lines about 0.1 inch apart, the curve of the vial being such that the travel of the bubble from one mark to the next represented a change in direction of the base, amounting to 0.012 inch per foot. By using a magnifying glass, much smaller variations than this could be detected. As an accurate straightedge of sufficient length was not available for testing the straightness of the V's, they were verified by using the spirit level and the blocks B, as shown by the right-hand view, Fig. 24. The blocks were first placed near one end of the casting, and the screw E adjusted until one end of the bubble coincided with one of the marks of the vial. The blocks and level were then moved to different positions along the length of the groove, and as the bubble remained in the same position, it was assumed that the planing was straight enough for the job in hand. In order to true up the V's of the lathe bed, a thin layer of Prussian blue was applied to the V-grooves of the templet which was then moved along the bed to transfer the coloring matter to the parts that needed scraping down. After a sufficient amount of scraping, the V's on the bed showed a bearing throughout their entire length when the templet was applied.

CHAPTER VIII

GAGES AND MEASURING INSTRUMENTS

Practically all of the measuring tools used by toolmakers and machinists, may be divided into two general classes; *viz.*, the tools for measurements of length, and those for the measurement of tapers or angles. In the two general classes of tools for length and angular measurements, there are many different types and designs. For instance, there is the adjustable type, which is graduated and is used for taking direct measurements in inches or degrees; then, there is another type which is fixed and cannot be used for determining various sizes or angles, but simply for gaging or testing one particular size. There are also tools for taking approximate measurements and others designed for very accurate or precise measurements. In this chapter the measuring and gaging tools designed more particularly for accurate work will be referred to.

The Vernier Caliper. — The vernier is an auxiliary scale that is attached to vernier calipers, height gages, depth gages, protractors, etc., for obtaining the fractional parts of the subdivisions of the true scale of the instrument. For example, the true or regular scale of the vernier caliper shown in Fig. 1, is graduated in fortieths of an inch, but by means of the vernier scale V, which is attached to the sliding jaw of the instrument, measurements within one-thousandth of an inch can be taken. In other words, the vernier, in this case, makes it possible to divide each fortieth of an inch on the true scale into twenty-five parts. The distance that the vernier scale zero has moved to the right of the zero mark on the true scale (which equals diameter D) may be read directly in thousandths of an inch, by calling each tenth on the true scale that has been passed by the vernier zero, one hundred thousandths, and each fortieth, twenty-five thousandths, and adding to this number as many thousandths as are indicated

by the vernier. The vernier zero in the illustration is slightly
beyond the five-tenths division; hence, the reading is 0.500 plus
the number of thousandths indicated by that line on the vernier
that exactly coincides with one on the scale which, in this case,
is line 15, making the reading 0.500 + 0.015 = 0.515 inch.

Principle of the Vernier Scale. — By referring to the enlarged
scales shown at A and B, Fig. 2, the principle of the vernier will
be more apparent. When a vernier caliper reads to thousandths
of an inch, each inch of the true scale S is divided into ten parts,
and each tenth into four parts, so that the finest divisions are

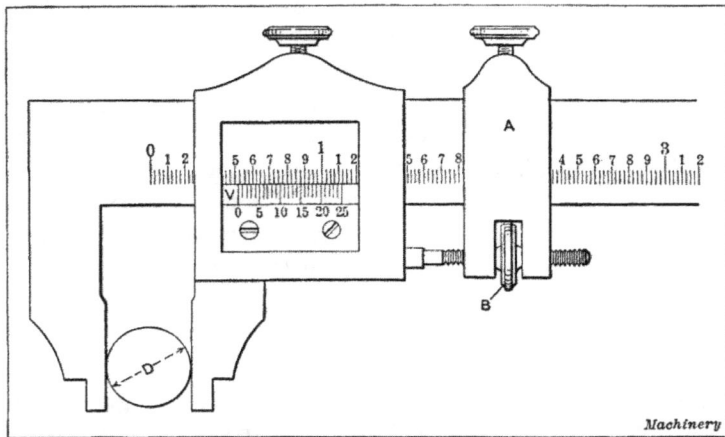

Fig. 1.   Vernier Caliper

fortieths of an inch. The vernier scale V has twenty-five divisions, and its total length is equal to twenty-four divisions on
the true scale, or $\frac{24}{40}$ of an inch; therefore, each division on the
vernier equals $\frac{1}{25}$ of $\frac{24}{40}$ or $\frac{24}{1000}$ inch. Now, as $\frac{1}{40}$ equals $\frac{25}{1000}$
we see that the vernier divisions are $\frac{1}{1000}$ inch shorter than those
on the true scale. Therefore if the zero marks of both scales
were exactly in line, the first two lines to the right would be $\frac{1}{1000}$
inch apart; the next two $\frac{2}{1000}$, etc. It is evident, then, that if
the vernier were moved to the right until, say, the tenth line
from the zero mark exactly coincides with one on the true scale,
as shown at A, the movement would be equal to 0.010 inch,
since this line was 0.010 inch to the left of the mark with which it

now coincides, when the zero lines of both scales were together. If the vernier were moved along to the position shown by the next sketch *B* (Fig. 2) the true scale would indicate directly that the reading was slightly over 0.500 inch, and the coincidence of the graduation line 15 on the vernier with a line on the true scale would show the exact reading to be 0.500 + 0.015 = 0.515 inch.

In Fig. 2 a true scale *S* is shown at *C* that is graduated into sixteenths of an inch, and the vernier *V* has eight divisions with a total length equal to seven divisions on the true scale, or $\frac{7}{16}$ of an inch; therefore, each division on the vernier is $\frac{1}{8}$ of $\frac{1}{16}$, or $\frac{1}{128}$ inch

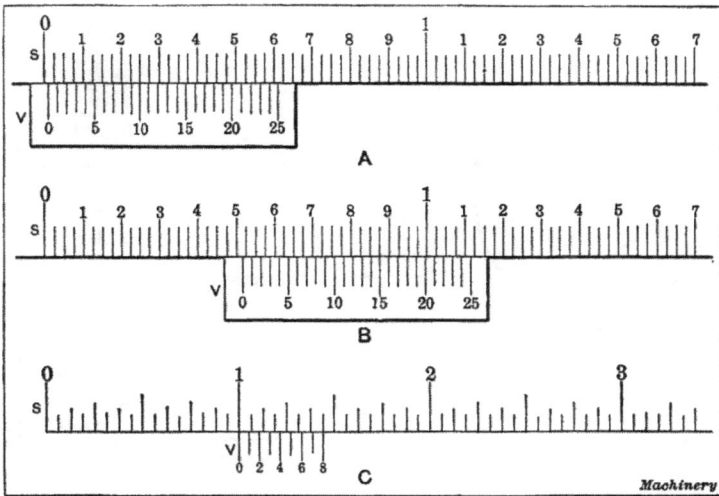

Fig. 2.   Scales with Verniers set in Different Positions

shorter than the divisions on the true scale; so we see that in this case the vernier enables readings to be taken within one hundred and twenty-eighths of an inch, instead of in thousandths as with the one previously described.

In order to determine the fractional part of an inch that may be obtained by any vernier, multiply the denominator of the finest subdivision of an inch given on the true scale by the total number of divisions on the vernier. For example, if (as in Fig. 1) the true scale is divided into fortieths and the vernier into twenty-five parts, the vernier will read to thousandths (40 × 25 = 1000).

If there are sixteen divisions to the inch on the true scale and a total of eight on the vernier, the latter will enable readings within one hundred twenty-eighths of an inch to be taken (16 × 8 = 128). It will be seen then that each subdivision on the true scale can be divided into as many parts as there are divisions on the vernier.

How to Read a Vernier. — The following is a general rule for taking readings with a vernier: *Note the number of inches and whole divisions of an inch that the vernier zero has moved along the true scale, and then add to this number as many thousandths, or hundredths, or whatever fractional part of an inch the vernier reads*

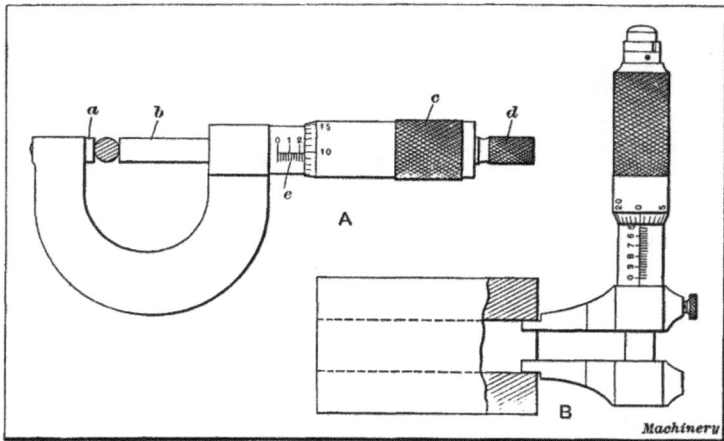

Fig. 3. Outside and Inside Micrometers

to, as there are spaces between the vernier zero and that line on it which coincides with one on the true scale.

Micrometers. — Micrometer calipers are indispensable in modern tool-rooms and machine shops for taking accurate measurements. A small size for external measurements is shown at *A*, Fig. 3. The part to be measured is placed against the anvil *a* and the adjustable spindle *b* is then screwed in until it bears lightly against the work, by turning the thimble or sleeve *c*; the size is then determined by referring to the micrometer graduations. Most micrometers are graduated to read to thousandths of an inch, although some have an auxiliary vernier scale which enables readings to within 0.0001 inch to be taken. Many

micrometers have what is called a ratchet stop *d* at the end of
the barrel or thimble.   If this is used when adjusting the measuring point against the work, it will slip when the point bears
lightly, and thus prevent excessive pressure.   The advantage of
securing a uniform contact or degree of pressure is that uniform
readings are then obtained.   Obviously, a difference in pressure
will give a different reading and might result in a serious error.
Inaccuracies from this cause might be negligible so far as
one workman is concerned, but they become important where
measurements are taken by many different workmen, because
everyone does not have the same sense of touch.

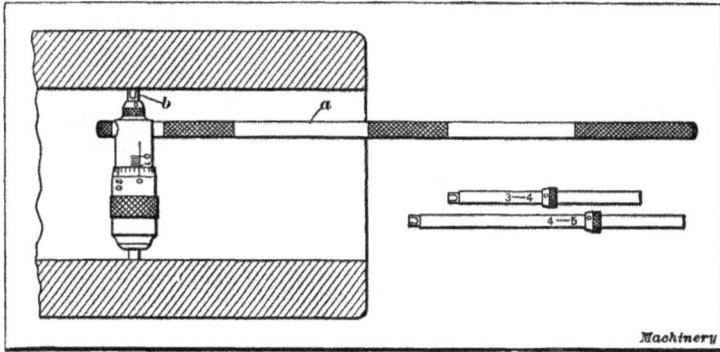

Fig. 4.   Inside Micrometer equipped with Extension Rods

A micrometer for measuring the diameters of holes or for taking
other internal dimensions is shown at *B*, Fig. 3.   The measuring
surfaces are hardened and ground to a radius to secure accurate
measurements and to avoid cramping when measuring the distances between parallel surfaces.   The movable jaw has a clamp
screw that is tightened when it is desired to retain the setting
of the calipers.   Another form of inside micrometer is shown in
Fig. 4.   This particular size can be used for measurements varying from 2 to 12 inches.   When testing the diameter of a comparatively small hole, when there is not sufficient room for the
hand, an auxiliary handle *a* is screwed into the micrometer head
as shown in the illustration.   The micrometer screw has a
movement of one-half inch and by inserting extension rods of

different lengths in the head at b, any dimension up to 12 inches can be obtained.   Two of these extension rods are shown to the right.   They are provided with collars which serve to locate them accurately in the micrometer head.

An inside micrometer gage that is especially adapted for large internal measurements is shown at A, Fig. 5.   This gage consists of a holder equipped with a micrometer screw with graduations reading to 0.001 inch, and into this holder is inserted an adjustable rod.   This rod also has graduations in the form of a series of annular grooves of a form and depth that allow clamp-

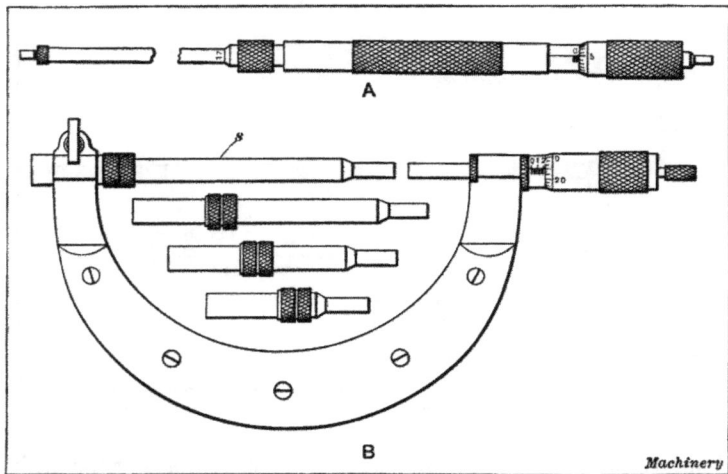

Fig. 5.   (A) Inside Micrometer Gage for Large Holes.   (B) Large Outside Micrometer

ing fingers on the holder to spring into them, thus making it possible to shift the rod in or out to the required length.   Gages of this type usually have a series of rods so that a wide range of sizes can be measured.   They are not only used for internal measurements but for setting calipers and for similar work.

A micrometer caliper for large external measurements is shown at B.   The micrometer screw has an adjustment of one inch and is graduated to read to 0.001 inch.   When measuring small sizes, the long anvil or spindle s is used, whereas, for larger sizes, one of the shorter spindles is inserted.   The sides of the steel frame

are covered with hard rubber to prevent inaccuracies in the measurements as the result of expansion from the heat of the hands.
As will be noted, this micrometer has a ratchet stop to insure
uniform pressure when measuring.

Thread Micrometers. — For the accurate measurement of
screws or threads, the special thread micrometer shown in Fig. 6
is often used. The fixed anvil is V-shaped so as to fit over the
thread, while the movable point is cone-shaped so that it will
enter the space between two threads. The contact points are on
the sides of the thread, as they must be in order that the pitch
diameter may be determined. The cone-shaped point of the
measuring screw is slightly rounded so that it will not bear at

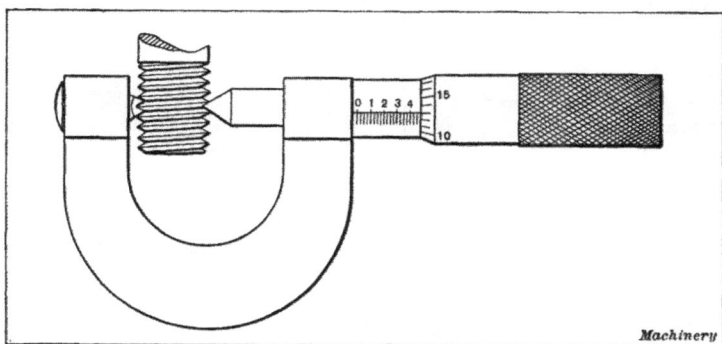

Fig. 6. Thread Micrometer

the bottom of the thread. There is also sufficient clearance at
the bottom of the V-shaped anvil to prevent it from bearing on the
top of the thread. The movable point is adapted to measuring
all pitches, but the fixed anvil is limited in its capacity. To
cover the whole range of pitches, from the finest to the coarsest,
a number of fixed anvils are required.

To find the theoretical pitch diameter, which is measured by
the micrometer, subtract the single depth of the thread from the
standard outside diameter. The depth of a V-thread equals
0.866 ÷ number of threads per inch, and depth of U. S. standard
thread equals 0.6495 ÷ number of threads per inch.

If standard plug gages are available, it is not necessary to
actually measure the pitch diameter, but merely to compare it

with the standard gage. In this case, a ball-point micrometer such as is shown in Fig. 7 may be employed. Two types of ball-point micrometers are ordinarily used. One is simply a regular micrometer with ball points made to slip over both measuring points, as shown by the detail sketch B. The best method is to use a regular micrometer into which ball points have been fitted as shown at A. A hole is provided in the spindle so that the ball point can easily be driven out when a larger or smaller size of ball point is required.

How to Read a Micrometer. — The pitch of the thread on the spindle b (Fig. 3) of an ordinary micrometer is $\frac{1}{40}$ of an inch. Along the frame at e (see also detail sketch A, Fig. 8), there are graduations which are $\frac{1}{40}$ inch apart; therefore, when thimble c

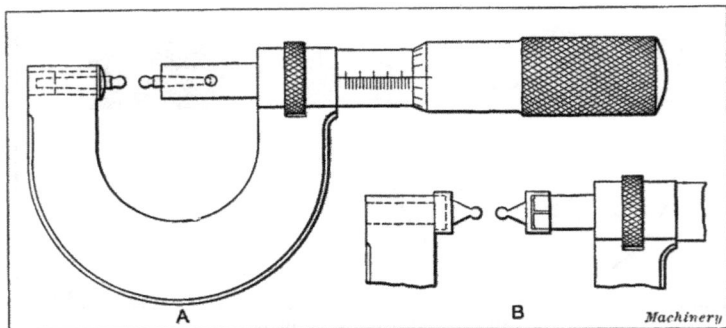

Fig. 7. Ball-point Thread Micrometer

and the measuring spindle are turned one complete revolution, they move in or out, a distance equal to one of the graduations or $\frac{1}{40}$ inch, which equals $\frac{25}{1000}$ inch. It is evident then that if instead of turning the thimble one complete revolution, it is turned say $\frac{1}{25}$ of a revolution, that the distance between the anvil and the end of the spindle will be increased or diminished $\frac{1}{25}$ of $\frac{25}{1000}$ of an inch, or one thousandth inch; therefore, the beveled edge of a micrometer spindle has twenty-five graduations, each of which represents 0.001 inch. Following is a general rule for reading a micrometer: *Count the number of whole divisions that are visible on the scale of the frame, multiply this number by 25 (the number of thousandths of an inch that each division represents) and add to the product the number of that division on the thimble which coincides with the axial zero line on the frame. The result will be the diameter expressed in thousandths of an inch.* As the numbers 1, 2, 3, etc., opposite every fourth subdivision on the frame indicate hundreds of thousandths, the reading can easily be taken mentally. Suppose the thimble were screwed out so that graduation 2, and three additional subdivisions were visible (as shown at *A*, Fig. 8), and that graduation 10 on the thimble coincided with the axial line on the frame. The reading then would be 0.200 + 0.075 + 0.010, or 0.285 inch.

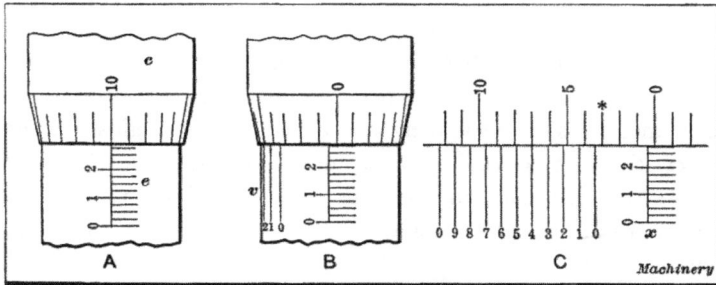

Fig. 8. Micrometer Graduations

Micrometer with Vernier Scale. — Some micrometers have a vernier scale *v* on the frame (see sketch *B*, Fig. 8) in addition to the regular graduations, so that measurements within 0.0001 inch can be taken. Micrometers of this type are read as follows: *First determine the number of thousandths, as with an ordinary micrometer, and then find a line on the vernier scale that exactly coincides with one on the thimble; the number of this line represents the number of ten-thousandths to be added to the number of thousandths obtained by the regular graduations.* The relation between the graduations of the vernier and those on the thimble is more clearly shown by diagram *C*. The vernier has ten divisions which occupy the same space as nine divisions on the thimble, and for convenience in reading are numbered as shown. The difference between the width of a vernier division and one on the thimble is equal to one-tenth of a space on the thimble. Therefore a movement of the thimble equal to this difference between the vernier and thimble graduations represents 0.0001 inch. When the thimble zero coincides with the line x on the frame, the zero of the vernier coincides with the third line to the left (marked with an asterisk). Now when the thimble zero (or any other graduation line on the thimble) has passed line x, the number of ten-thousandths to add to the regular reading is equal to the number of that line on the vernier which exactly coincides with a line on the thimble. Thus the reading shown at C (Fig. 8) is 0.275 + 0.0004 = 0.2754 inch.

Fixed Gages. — While any tool or instrument used for taking measurements might properly be called a gage, this term, as used by machinists and toolmakers, is generally understood to mean that class of tools which conform to a fixed dimension and are used for testing sizes but are not provided with graduated adjustable members for measuring various lengths or angles. There are exceptions, however, to this general classification. Measuring instruments, such as the micrometer and vernier caliper, are indispensable because they can be used for determining actual dimensions, and, being adjustable, cover quite a range of sizes. Any form of adjustable measuring tool, however, has certain disadvantages for such work as testing the sizes of duplicate parts, especially when such tests must be made repeatedly, and solid or fixed gages are commonly used. There is less chance of inaccuracy with a fixed gage and it is more convenient to use than a tool which must be adjusted, but owing to the necessity of having one gage for each variation in size, and because of the cost of a set covering a wide range of sizes, solid gages are used more particularly for testing large numbers of duplicate parts in connection with interchangeable manufacture.

Two different types of fixed gages are shown in Fig. 9. The form shown at A is commonly known as a " snap gage." The distance between the measuring surfaces is fixed and represents the size stamped upon the gage, within very close limits. This type of gage can be obtained in various sizes and is used for measuring duplicate parts in connection with general shop work. Sketch B illustrates another form of snap or caliper gage. This is double-ended and is intended for both external and internal measurements, the width of the internal end being the same as the distance between the measuring surfaces of the external end.

Limit Gages. — With the modern system of interchangeable manufacture, machine parts are made to a definite size within certain limits which are varied according to the accuracy required, which, in turn, depends upon the nature of the work. In order to insure having all parts of a given size or class, within the prescribed limit so that they can readily be assembled without extra and unnecessary fitting, what are known as " limit gages " are used. One form of limit gage for external measurement is

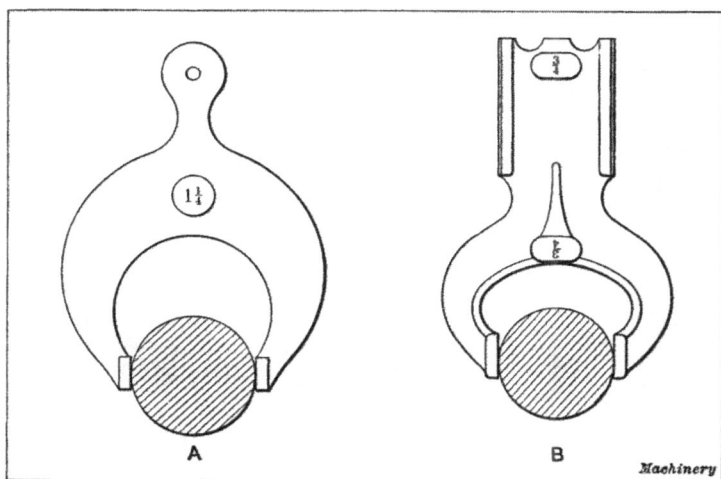

Fig. 9. (A) Snap Gage. (B) Combined Internal and External Gage

shown at *A*, Fig. 10. It is double-ended and has a "go" end and a " not go " end; that is, when the work is reduced to the correct size, one end of the gage will pass over it but not the other end. When a single-ended snap gage *A*, Fig. 9, is used, the diameter of the work may be slightly less than it should be, but by having a gage for the minimum as well as for the maximum size, every part must come within the limits of the gage. This allowance or limit is made to conform to whatever amount experience has shown to be correct for the particular class of fit required. Another external limit gage is shown at *B*, Fig. 10. Nominally this is a $\frac{1}{4}$-inch gage. The size of the " go " end is 0.250 inch and the size of the " not go " end is 0.2485 inch; hence the tolerance is 0.0015 inch. Therefore a part that is more than 0.0015 inch less than 0.250 inch will pass the " not go " end of the gage. An internal limit gage is shown at C. The nominal size of this particular gage is $1\frac{1}{4}$ inch. The diameter of the " go " end is 1.2492 inch, whereas the diameter of the " not go " end is 1.2506 inch; hence, in this case, the tolerance equals 1.2506 − 1.2492 = 0.0014 inch. Incidentally, it is good practice to make all holes to standard sizes within whatever limits may be advisable, and vary the size of the cylindrical parts to secure either a forced fit,

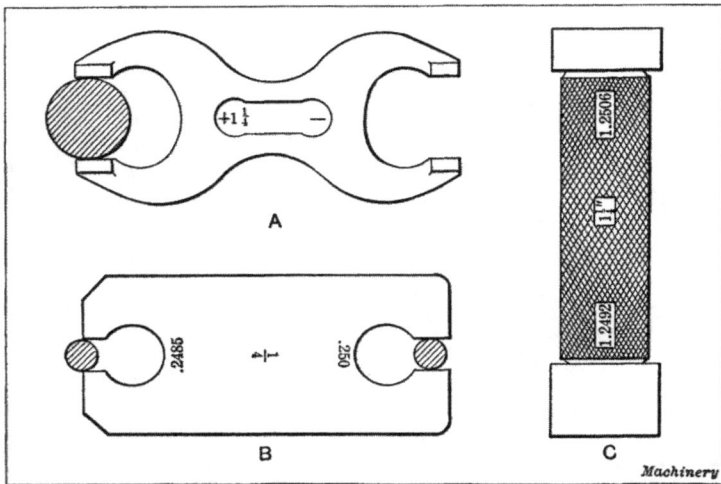

Fig. 10. External and Internal Limit Gages

running fit, or whatever class of fit may be required. It will be noted that the ends of these limit gages are of different shape so that the large and small sizes can readily be identified without referring to the dimension stamped on the gage ends. Limit gages are very generally used for the final inspection of machine parts, as well as for testing sizes during the machining process. They are superior to the micrometer for many classes of inspection work, because the adjustment and reading necessary with a micrometer often results in slight variations of measurement, especially when the readings are taken by different workmen.

Adjustable Limit Snap Gage. — The snap gage shown at A, Fig. 11, differs from the ordinary single-ended type in two particulars: In the first place, it has two sets of measuring plugs and is a limit gage. The lower set forms the " go " end and the upper set the " not go " end. These plugs are also adjustable so that when the gage becomes inaccurate, as the result of wear, the plugs can easily be reset, a standard reference gage being used to determine the distance between them. The plugs are plain cylinders of hardened steel and are lapped to a snug sliding fit in the hole of the gage body. The ends are square and bear against adjusting screws, the forward ends of which are also lapped

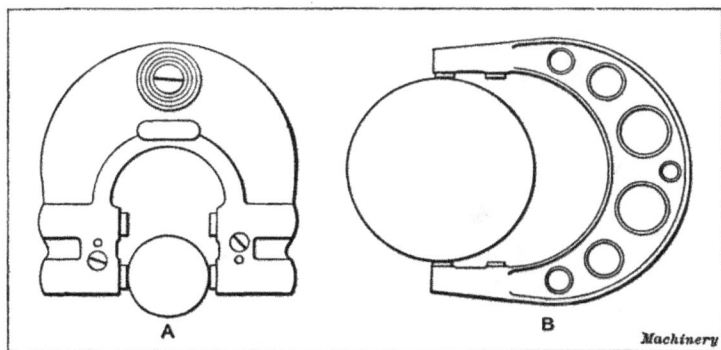

Fig. 11. (A) Adjustable Limit Gage. (B) Limit Gage with Fixed Points

square. The clamping screws at the side not only clamp the plugs but tend to force them against the adjusting screws. The handle has an insulated grip. Another snap gage of the limit type is shown at B. This gage has fixed points which can be renewed in case of wear.

Plug and Ring Gages. — A standard external or ring gage and internal or plug gage is shown at A, Fig. 12. These gages are very accurately made and are used either as reference gages or for setting calipers, etc., or as working gages. One gage manufacturer makes solid gages of this type in diameters varying from $\frac{1}{16}$ inch to 3 inches. For larger sizes, up to 6 inches in diameter, the plug gages are made hollow. U. S. standard thread gages are shown at B, Fig. 12. These gages are intended as a practical working standard. The internal gage or plug is a standard to which the external templet is adjusted. The plain unthreaded end of the plug gage is ground and lapped to the exact diameter at the root or bottom of the thread. Gages for testing the accuracy of tapers are also made in the form of a plug and ring (as at *A*, Fig. 12) excepting, of course, that they are made tapering. The ring gage is used for external tapers and the plug for holes. The plug accurately fits the ring and when they are assembled, a line on the plug coincides with the end of the ring. This line is used for gaging the depth of holes which must conform to the standard size of the ring gage. When the plug gage is used as a working gage in the shop, the ring is usually kept as a

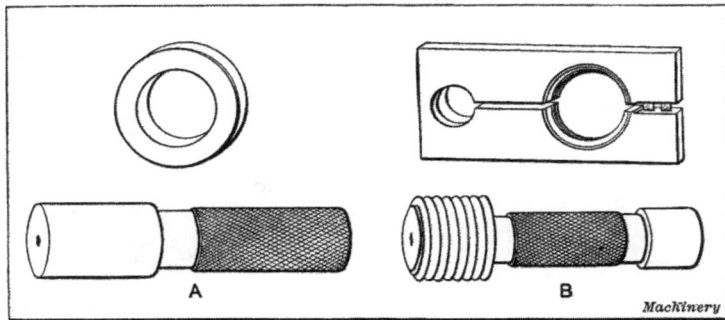

Fig. 12. (A) Plug and Ring Gages. (B) Internal and External
Thread Gages

reference gage. On the other hand, if a ring is used for testing external tapers, the plug is often preserved as the reference gage.

Gage for Accurately Measuring Tapers. — When a certain taper or angle must be originated or accurately measured, the disk type of gage shown in Fig. 13 may be employed. This gage consists of two adjustable straightedges A and A_1, which are in contact with disks B and B_1. The angle α or the taper between the straightedges depends, of course, upon the diameters of the disks and the center distance C, and as these three dimensions can be measured accurately, it is possible to set the gage to a given angle within very close limits. Moreover, if a record of the three dimensions is kept, the exact setting of the gage can easily be reproduced at any time. The following rules may be used for adjusting a gage of this type.

To Find Center Distance for a Given Angle.— When the straight-edges must be set to a given angle α, to determine center distance C between disks of known diameter. *Rule:* Find the sine of half the angle α in a table of sines; divide the difference between the disk diameters by double this sine.

To Find Center Distance for a Given Taper. — When the taper, in inches per foot, is given, to determine center distance C. *Rule:* Divide the taper by 24 and find the angle corresponding to the quotient in a table of tangents; then find the sine corresponding

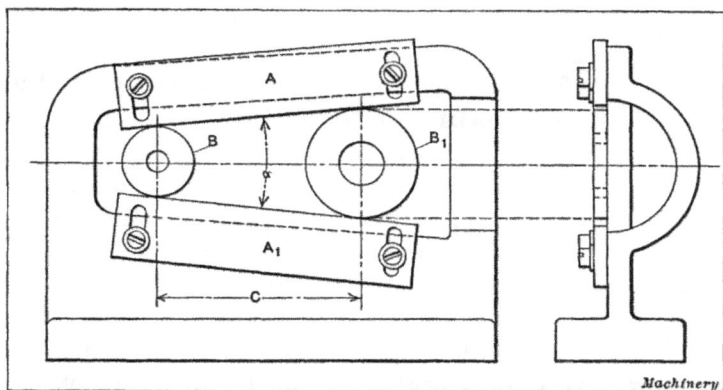

Fig. 13. Disk Gage for Originating or Accurately Measuring Tapers or Angles

to this angle and divide the difference between the disk diameters by twice the sine.

To Find Angle for Given Disk Dimensions. — When the diameters of the large and small disks and the center distance are given, to determine the angle α. *Rule:* Divide the difference between the disk diameters by twice the center distance; find the angle corresponding to the quotient, in a table of sines, and double the angle.

To Find the Taper per Foot. — When the diameters of the large and small disks and the center distance C are given, to determine the taper per foot (measured at right angles to a line through disk centers). *Rule:* Divide the difference between the disk diam-

eters by twice the center distance; find the angle corresponding
to the quotient, in a table of sines; then find the tangent corresponding to this angle, and multiply the tangent by 24.

Reference Gages. — Reference gages are intended for testing
the accuracy of working gages such as are used in the shop and
tool-room, and for setting other forms of measuring instruments.
Reference gages are made in different forms, varying from plain
blocks or disks to special shapes designed for some particular
class of work. Plug and ring gages similar to the type illustrated
at *A*, Fig. 12, are also used to some extent for reference purposes,
as well as for working gages. In some shops it is the practice to
use the plug as a working gage and the ring for testing it, or, in
case the ring is required as a working gage, the plug is kept as a
standard or reference gage, as previously mentioned. End-
measuring rods and blocks are often used for testing snap gages,
etc. Ordinarily, the solid measuring rods are cylindrical in form
and may be obtained in sets covering a considerable range of
lengths. These rods are used for testing the parallelism and
width of two finished surfaces, as well as for setting calipers and
testing gages. The ends of some rods are made flat and parallel,
whereas others have ends which are sections of spheres, the diameters of which equal the lengths of the rods. The spherical-
ended form is very convenient for testing the diameters of rings,
cylinders, etc. Some end-measuring rods are provided with an
insulating handle in the center to prevent expansion from the
heat of the hand.

Johansson or Swedish Gages. — The Johansson combination
standard gages consist of a series of rectangular steel blocks which
are finished on all sides with wonderful accuracy. The opposite
sides of each block are parallel and the distance between them is
equal to the dimension stamped upon the block, within a limit so
small as to be inconceivable. The eighty-one blocks in what is
known as Set No. 1 (see Fig. 14) are arranged in four series. The
first series contains 9 blocks which vary in thickness from 0.1001
inch to 0.1009 inch, increasing by 0.0001 inch increments. The
second series contains 49 blocks, varying in thickness from 0.101
inch to 0.149 inch, increasing by 0.001 inch. In the third series

there are 19 blocks, varying in thickness from 0.050 inch to 0.950 inch, increasing by 0.050 inch.   The last series of four blocks has 1, 2, 3 and 4 inch sizes, respectively.   The gages for the English system of measurement are adjusted to their sizes at 66° F.   The value of these gages lies in the fact that they are not only exceptionally accurate, but are so varied in size that, with the set referred to in the foregoing, a gage 10 inches long can be built and dimensions varying by 0.0001 inch be obtained.   According to the makers, this one set will give at least 100,000 gage sizes, by using the various combinations of blocks which are possible.

Fig. 14.   Johansson Reference Gages

Any dimension up to 8 inches obtained by the systematic combination of these blocks is said to be exact within 0.00004 inch; hence, the error of any one block is exceedingly small.

How to use Johansson Gages. — The combination of these Johansson gages to form any required dimension is simple but should be done systematically.   Every block is marked with its size and in placing two blocks together they are slid over each other with a slight pressure.   Any dust that might be on the surfaces should first be removed by using the finger.   To illustrate how the gages are combined, suppose 3.4566 inches is the required size.   First it is well to consider the ten-thousandths in the dimension; therefore, block 0.1006 (which is one of the first series previously mentioned) would be selected. The thousandths in the dimension are next taken care of by selecting block 0.106. The block for the even number of inches, or the 3-inch size, is then added, which makes the dimension 3.2066 inches; therefore, the block needed to complete the dimension is 0.250. Thus, the entire set is made up as follows:

$$0.1006 + 0.106 + 3 + 0.250 = 3.4566 \text{ inches.}$$

This same dimension could also be obtained by using an entirely different combination. In order to show how different combinations can be used for obtaining the same size, suppose the dimension 0.600 inch is required. Gages of this size could be made up by using the following combination: 0.550 + 0.050; 0.450 + 0.150; 0.400 + 0.200; 0.350 + 0.250; 0.500 + 0.100, etc.

If a $1\frac{5}{8}$-inch gage were required, the 1 inch, 0.500 inch and 0.125 inch blocks could be used. Thus: $1 + 0.500 + 0.125 = 1.625$ or $1\frac{5}{8}$ inch. If a size 0.002 inch larger or 1.627 inch were required, this could be obtained simply by substituting the 0.127 inch block for the 0.125 inch size. Other combinations could also be used for the size given in the preceding example. From the foregoing, it will be seen that a gage can be built up which will include the plus allowance for a forced fit, the minus allowance for a running fit, or any tolerance or limit which may be desired.

Adhesion of Johansson Gages. — A remarkable property of the Johansson or so-called Swedish gages is their adhesiveness to one another. When wrung together they will resist separation in a direction at right angles to the faces in contact, with a force considerably greater than the atmospheric pressure on the area of contact. This phenomenon has caused some to believe that actual molecular adhesion takes place when surfaces that are nearly perfect planes are brought into intimate contact. Tyndall, the eminent English physicist, reached the conclusion thirty-five years ago that molecular attraction was partly the cause of the phenomenon. The error of this theory has been shown by some investigations reported to the London Royal Society, showing that the adhesion results from the presence of a

very thin liquid film. Some blocks of hardened steel were prepared, each weighing an ounce and a half and having surfaces of
0.7 square inch polished flat to within a millionth of an inch of
accuracy, and these were used to test the adhesive properties
of many liquids. The contact faces were carefully freed from
moisture and grease with alcohol before being smeared with a
very thin film of the liquid under test. When these blocks were
wrung together while the surfaces were perfectly clean, they fell
apart under their own weight; but blocks held together by films
required to separate them, a force ranging from seventeen pounds
for Rangoon oil, to twenty-two for lubricating oil, twenty-nine

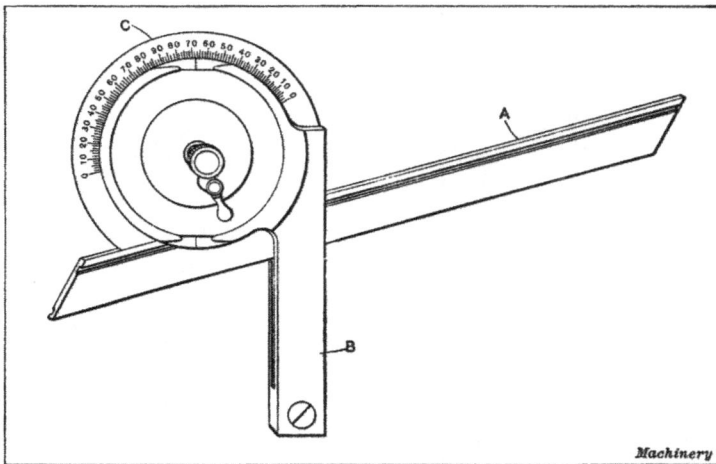

Fig. 15. Universal Bevel Protractor

for turpentine, and thirty-five for condensed water vapor. After
washing the hands with soap, blocks rubbed on them showed
adhesion as high as ninety pounds. There was no adhesion
from volatile liquids, like alcohol and benzine, and very little
from viscous liquids, such as glycerine and glucose. The microscope showed that the films, drawn out in thin lines, covered only
a tenth or less of the metal faces. From varied experiments it
appeared that in the case of paraffine film, for instance, the
twenty-seven pounds required to part the plates included about
one pound due to atmospheric pressure, one to surface tension

and twenty-five pounds to the actual tensile strength of the liquid.

The Protractor. — There are many different forms of protractors, but they all embody the same general principle. The type commonly used by machinists and toolmakers has a straight-edge or blade which can be set at any angle with the base or stock, and the angle for any position is shown by degree graduations. This form is generally known as a bevel protractor. A common design of bevel protractor is shown in Fig. 15. The angular position between blade *A* and stock *B* can be varied as may be required, and disk *C*, which is graduated from 0 to 90 degrees in each direction, shows what the angle is for any position. The

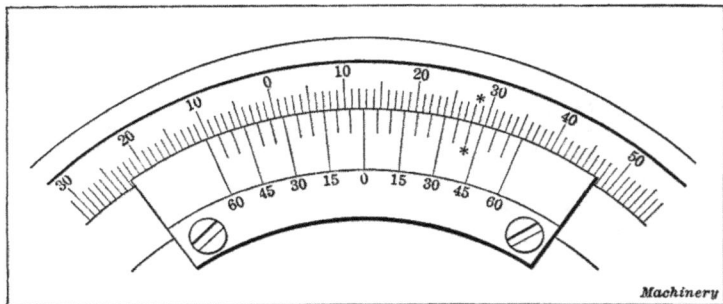

Fig. 16.   Protractor Scale and Vernier

blade, which is clamped by an eccentric stud, can be adjusted in a lengthwise direction so that it can be used in any position.

Reading a Protractor Vernier. — When the graduations on protractors are not finer than whole degrees, measurements of fractional parts of a degree cannot be made with accuracy but by the addition of a vernier scale, subdivisions of a degree are easily read. The vernier scale of a universal bevel protractor is shown in Fig. 16. This particular vernier makes it possible to determine the angle to which the instrument is set, within five minutes (5′) or one-twelfth of a degree. It will be noted that there are practically two scales of twelve divisions each, on either side of the vernier zero mark. The left-hand scale is used when the vernier zero is moved to the left of the zero of the true scale, while the right-hand scale is used when the movement is to the right.   The total length of each of these vernier scales is equal to twenty-three degrees on the true scale, and as there are twelve divisions, each division equals $\frac{1}{12}$ of 23 or $1\frac{11}{12}$ degree.   One degree equals 60 minutes (60′), and $\frac{11}{12}$ degree equals $\frac{11}{12}$ of 60 or 55 minutes; hence each division on the vernier expressed in minutes equals 60′ + 55′ = 115 minutes.   Now as there are 120 minutes in 2 degrees, we see that each space on the vernier is 5 minutes shorter than 2 degrees; therefore, when the zero marks on the true and vernier scales are exactly in line, the first graduation (either to the right or left) on the vernier is 5 minutes from the first degree graduation; the next two are 10 minutes apart; and the next two 15 minutes, etc.   It is evident then that if the vernier is moved, say to the right, until the third line from zero is exactly in line with one on the true scale, the movement will be equal to 15 minutes, as indicated by the number opposite this line on the vernier.

To read the protractor, first note the number of whole degrees passed by the vernier zero, and then count in the same direction the number of spaces between the vernier zero and that line which exactly coincides with one on the regular scale; this number of spaces multiplied by 5 will give the number of minutes to be added to the whole number of degrees.   The reading of a protractor set as illustrated in Fig. 16 is 12 whole degrees plus 40 minutes.   The vernier zero has passed the twelfth graduation and the eighth line on the vernier coincides with a line on the true scale; hence 40 minutes is added to 12 degrees to get the correct reading.

Sine-bar for Measuring Angles. — The sine-bar is used either for measuring angles accurately or for locating work to a given angle.   It consists of an accurate straightedge to which are attached two hardened and ground plugs p and p_1 (see Fig. 17).   These plugs must be of the same diameter, and the distance l between their centers should, preferably, be an even dimension, to facilitate calculations.   The edges of the straightedge must be parallel with a line through the plug centers.   The sine-bar is always used in conjunction with some true surface B from which measurements can be taken.   Two methods of measuring an

angle are illustrated. Referring to the left-hand sketch, the
upper edge A of the part to be measured is set parallel with surface plate B. The heights a and b from the surface plate to the
plugs p and p_1 are carefully measured either by using a micrometer gage or a vernier height gage. The difference between a
and b is determined, and this difference, divided by the length l
between the plugs of the sine-bar, equals the sine of the required
angle β. The angle is then found by referring to a table of sines.
For example, suppose length l is 10 inches, height a, 7.256 inches
and height b, 2.14 inches; then the sine of the required angle
equals $(7.256 - 2.14) \div 10 = 0.5116$, which is the sine of 30
degrees 46 minutes. A 10-inch sine-bar is convenient to use, as

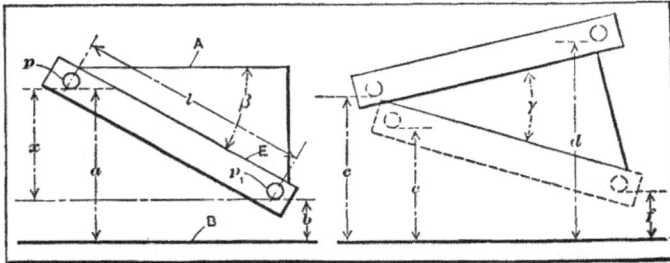

Fig. 17.   Diagrams Showing how Sine-bar is used for Measuring
Angles

division can be performed mentally by simply moving the deci-
mal one point to the left. Fig. 18 illustrates how the sine-bar A
is used to determine the angle between the lower edge of triangle
B and the machine table. A micrometer gage is used for measuring the vertical heights of the plugs.

The sketch to the right in Fig. 17 illustrates a method of measuring an angle without first setting one edge parallel to surface
B, the angle of each edge being measured separately. Suppose
the height d equals 8.75 inches and c equals 6.5 inches. Subtracting c from d: $8.75 - 6.5 = 2.25$. Next shift the sine-bar
to the position shown by the dotted lines. Assuming that $e = 5$
inches and $f = 2.15$, then $e - f = 5 - 2.15 = 2.85$. Dividing
2.25 and 2.85 by 10 (or the center distance between the sine-bar
plugs) we get 0.225 and 0.285 as the sines of the angles; 0.225 is

the sine of 13 degrees 1 minute, and 0.285 is the sine of 16 degrees 34 minutes. The sum of these angles or $(13° 1') + (16° 34') = 29$ degrees 35 minutes or the required angle γ.

When the sine-bar is to be set to a given angle for locating some part with reference to it, it is first set approximately. The sine of the required angle is then found and this sine is multiplied by the distance l between the plug centers, to obtain the vertical

Fig. 18. Setting Sine-bar with Micrometer Gage

distance x (see left-hand sketch, Fig. 17) for that particular angle. The bar is then adjusted until the vertical distance x coincides with the dimension found. For example, if edge A is to be ground to an angle of 30 degrees 46 minutes from edge E, the sine-bar is clamped to the angle-plate at approximately this angle. The sine of 30 degrees 46 minutes, or 0.5116, is then multiplied by 10 to obtain the vertical distance x, and the bar is adjusted by the use of a vernier height gage until x equals 0.5116×10 = 5.116 inches.

Testing Accuracy of Squares. — Two methods of testing the accuracy of a try-square are shown in Fig. 19. In order to make a reliable test, a 90-degree angle should be originated, unless a master square of known accuracy is available. A comparatively simple way of doing this accurately is to make a cylindrical plug similar to the one shown at A. The lower end of this plug is recessed to form a narrow edge which is beveled on the outside so that there will be no bearing in the corner where the blade joins the stock. This plug is ground on dead centers and

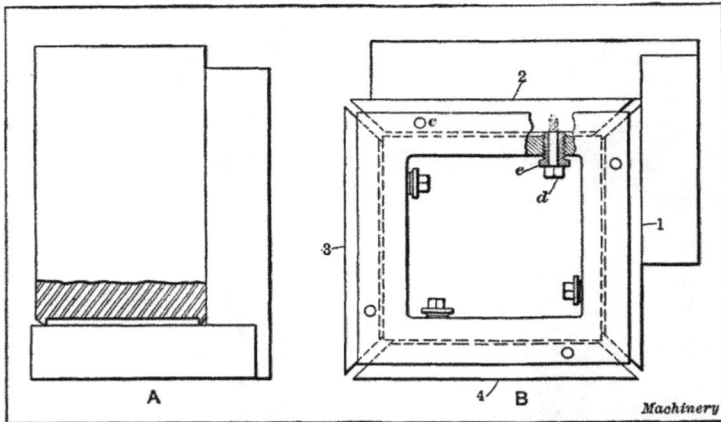

Fig. 19. Two Methods of Testing a Square

lapped to form as perfect a cylinder as possible. The narrow edge at the end is then ground true so that it will be exactly at right angles to the cylindrical surface. By holding the square against the side and end of the plug, as the illustration indicates, and subjecting it to the light test, a very minute inaccuracy in the position of the square blade can be detected. The outside edge of the blade can be tested by placing the plug and square on an accurate surface plate, and bringing the blade edge into contact with the side of the plug.

Precision Test Block for Squares. — A more elaborate form of test block than the one previously referred to, which gives very accurate results, is shown at B, Fig. 19. This test block is formed of a square cast-iron frame which is grooved around the outside and contains four close-fitting adjustable strips which, in the illustration, are numbered from 1 to 4. The reliability of this test block depends largely upon the outer edges of these strips which must be accurately finished plane surfaces. The strips are held in place by close-fitting pins c near the ends, and by bolts d. The latter pass through clearance holes in set-screws e which are screwed through the frame and bear against the inner edges of the strips. By clamping one of these strips against the set-screws, it is locked in position after being properly adjusted.

The method of using this test block for determining the accuracy of a try-square is as follows, assuming that the edges have not previously been adjusted: The square is first placed against two of the strips or straightedges of the test block. These strips are then adjusted until they exactly fit the square being tested. If the square were first applied to strips Nos. 1 and 2 (as shown in the illustration) strips 2 and 3 would next be set in the same manner, and then strips 3 and 4. After making these adjustments, if the square is applied to the strips Nos. 4 and 1, any error which might exist would be multiplied four times; whereas, if the square fitted these last sides perfectly, this would indicate that the angle between the square blade and stock was 90 degrees, within very close limits.

To illustrate how the error accumulates in going around the test block, suppose the angle between the blade of a square and its stock were 90 degrees 15 minutes. Evidently, then, sides 1 and 2 of the test block would also be set to this angle. Therefore, taking side No. 1 as a base, side No. 2 would be out 15 minutes. As side 2 is used in setting side 3, the error of the latter with reference to side 1 would be 30 minutes; similarly, side 4 would have an error of 45 minutes, and when the square was applied to sides 4 and 1 for the final test, the error would be four times the original amount, or 1 degree.

In order to originate a 90-degree angle, or, in other words, to set the test block to this angle, a sheet steel templet may be used. This simply forms a temporary try-square and is cut away so that there are two small projections along each test edge, in order that changes can be made by simply altering these small projections. This templet is first made as accurately as possible and it is then used in setting the test block. After adjusting the block, if comparison with the fourth and first sides shows an error, the templet is corrected and the test block again adjusted. This operation is repeated until the 90-degree angle is originated. The accuracy of a square can then be tested by comparison with any two sides of the test block and without making any adjustments.

Disk Method of Testing Squares. — Fig. 20 shows a method of testing and adjusting try-squares which requires four disks

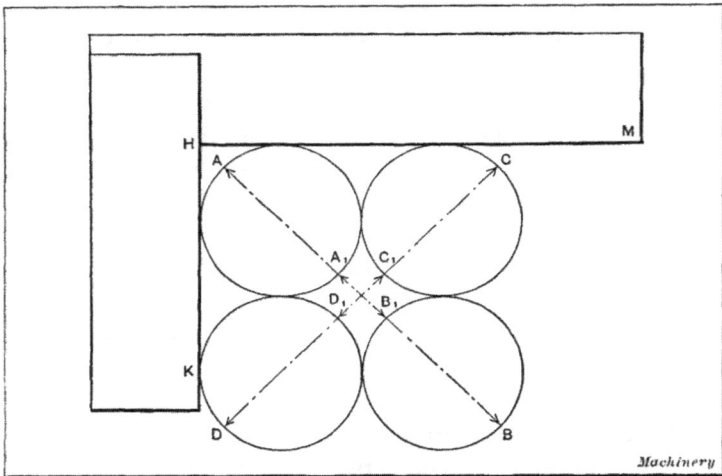

Fig. 20. Testing a Square by Means of Disks

of exactly the same size, the size being immaterial except that the disks should, of course, be in proportion to the size of the square to be tested. To obtain four disks of the same size is not difficult, that is, within limits of, say, one-ten-thousandth of an inch, which is perhaps closer to absolute accuracy than the average try-square will ever be required. Having four such disks, arrange them as shown in the engraving, and then, with outside calipers, test distances AB and CD (or with inside calipers test distances A_1B_1 and C_1D_1). If the two distances measure the same, the line KH and HM must form a right angle. Any deviation from a right angle will be multiplied by two, making this test extremely accurate. Thus if the disks crossed by the line indicating the distance AB are closer together by one-thousandth of an inch than they should be for the right-angle position, the other two disks will be pushed one-thousandth of an inch apart, making the difference in the measurements two-thousandths. This test does not, of course, prove anything except the position of the points touched by the disks, but in practical try-square making the first thing to do is to get the blade and beam as nearly perfect straightedges as possible before fastening them together, and they should be so tested before applying a test for the right-angle position.

Straightedges. — Straightedges are used to test flat surfaces for determining whether or not they are true planes, and also for testing round parts for bends, or curvatures in a lengthwise direction. Perhaps the most common form of straightedge is of rectangular section. In order to increase the sensitiveness of a straightedge for showing minute deviations or curvatures, the testing edge is made narrower by beveling one side, thus decreasing the width to about $\frac{1}{16}$ inch. For work requiring extreme accuracy, the testing edge should be very narrow and of semi-circular cross-section so that a line contact is obtained instead of a surface contact. This line contact shows any minute curvature which may exist and as the edge is curved the accuracy of the test will not be affected if the straightedge is not held exactly at right angles to the surface being tested. When using a straightedge having plane or flat surfaces, it should be held square with the work, because, if canted so that only one edge is in contact, any inaccuracy along this edge would appear as an inaccuracy in the surface being tested. When comparing a surface with a straightedge, there should be a good light on the side opposite the observer so that any irregularities or curvatures in the work can readily be detected.

Height and Depth Gages. — The vernier height gage, shown at A, Fig. 21, is used for locating jig buttons, measuring the vertical distance from one plane surface to another, etc. It is similar to a vernier caliper, except that there is a rather heavy base which allows the gage to stand upright. The movable jaw of this particular make of gage has a projection which extends beyond the base and is convenient for testing the height of a button attached to a jig plate (as the illustration indicates) and for similar work. The end of this extension is beveled to a sharp edge for scribing lines. The gage is graduated to read to thousandths, by means of a vernier scale on the sliding jaw. There are graduations on both sides, giving readings on one side for outside measurements and on the other side for inside measurements. This particular gage can be used for heights up to 8 inches.

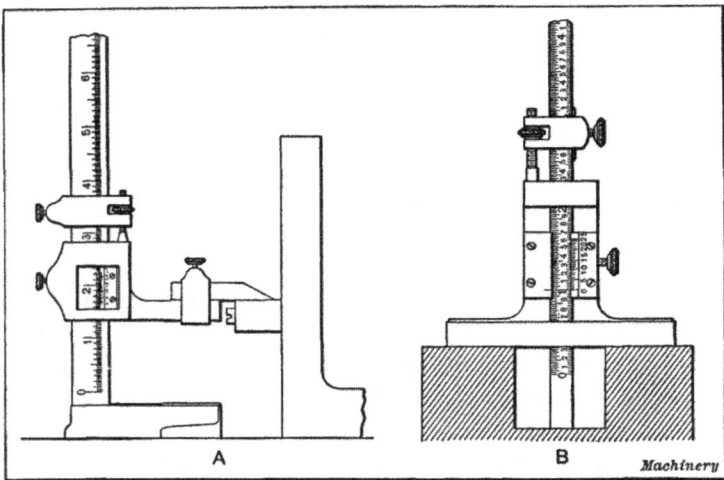

Fig. 21.   (A) Vernier Height Gage.   (B) Vernier Depth Gage

Illustration B, Fig. 21, shows a depth gage for measuring the depths of holes, recesses in dies, etc. The vertical blade or scale is graduated and by means of a vernier gives readings to thousandths of an inch. Height and depth gages are also made on the micrometer principle; that is, instead of having a scale and vernier, the adjustments are effected by a micrometer screw, graduated to read to thousandths.

Center Indicator. — The center indicator is used to set any point or punch mark in line with the axis of a lathe spindle preparatory to boring a hole. The plan view, Fig. 22, shows how the indicator is used. It has a pointer A, the end of which is conical and enters the punch mark to be centered. This pointer is held by shank *B* which is fastened in the toolpost of the lathe. The joint *C*, by means of which the pointer is held to the shank, is universal; that is, it allows the pointer to move in any direction. When the part being tested is rotated by running the lathe, if the center punch mark is not in line with the axis of the lathe spindle, obviously, the outer end of pointer *A* will vibrate, and as the joint *C* is quite close to the inner end, a very slight error in the location of the center punch mark will cause a perceptible movement of the outer end, as indicated by the dotted lines. Obviously, when the work has been adjusted until the pointer

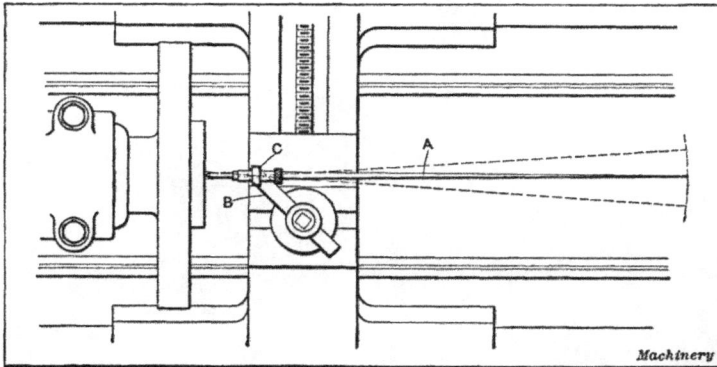

Fig. 22. Plan View Illustrating Use of Center Indicator

remains practically stationary, the punch mark is in line with the axis of the lathe spindle. When two holes are being bored to a given center-to-center distance, by first laying out the centers and then indicating them true in this way, the accuracy depends largely upon the location of the center punch marks.

Test Indicators. — The test indicator is extensively used in connection with the erection of machinery, for detecting any lack of parallelism between surfaces, in inspection departments, and for testing the accuracy of rotating parts such as spindles or arbors. Fig. 23 shows how a dial indicator is used to test the concentricity of the outer race of a roller bearing. The assembled bearing is mounted upon an accurately running arbor, held between centers, and the contact point *A* of the indicator bears

against the surface of the outer race.   As the latter revolves, the
slightest deviation or eccentricity is shown by vibrations of the
dial hand, which is so connected with the contact point that any
motion of the latter is magnified a number of times.   The graduations on the dial face indicate thousandths of an inch, and the
dial is adjustable so that it can be turned to locate the zero mark
directly under the hand, after the contact point has been adjusted
against the work.   The graduations then give a direct reading in
thousandths for any deviation from the central or zero position.

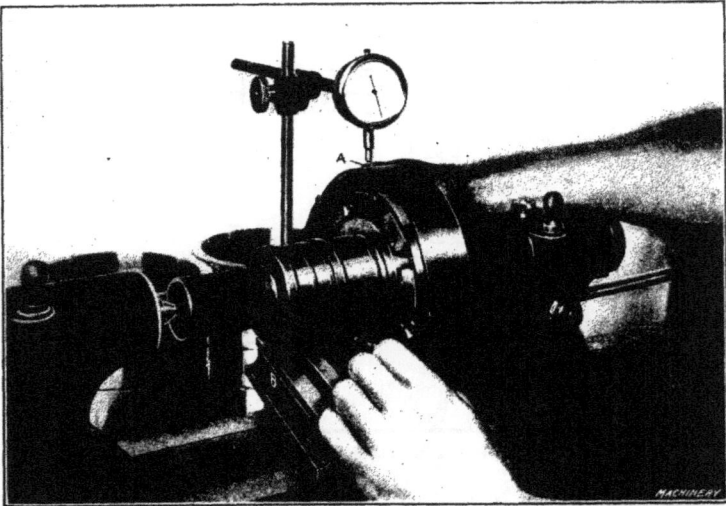

Fig. 23.   Testing Concentricity of Roller Bearing with Dial Test
Indicator

The contact point is removable to permit inserting different
forms.   In this particular case, the indicator is supported by a
vertical rod attached to a base B, which forms part of the instrument.   It is often used independently of the base, as when held
in the toolpost of a lathe for testing the accuracy or concentricity
of a cylindrical surface.   The dial indicator is also used for many
other purposes.   For instance, it is often attached to a surface
gage, in place of the pointer or scriber, for testing the parallelism
of a surface, especially when it is desirable to know the exact
amount of inaccuracy.

Two other forms of test indicators are shown in Fig. 24. This type is also used in connection with the erection or inspection of machinery for detecting inaccuracies, such as the lack of parallelism between two surfaces or the amount a cylindrical part runs out of true. Diagram *A* illustrates how a jig button is set true with the lathe spindle. The point of the indicator is set against the button and, as the latter revolves, any inaccuracy is shown by the vibrations of the pointer. Any movement of the contact

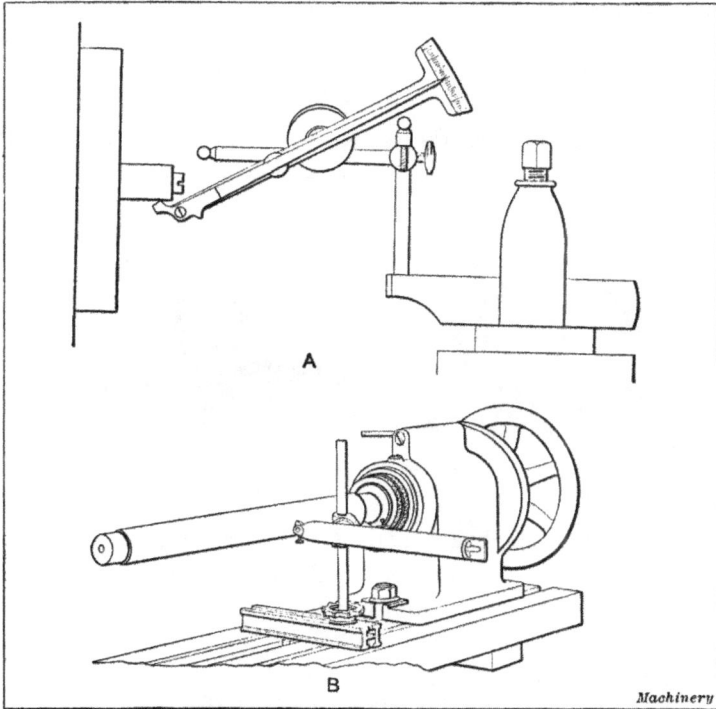

Fig. 24. Examples Illustrating Use of Test Indicators

point is multiplied several times by the pointer, and graduations at the end of the latter indicate thousandths of an inch. Diagram *B* illustrates how another test indicator of different form is used for determining the accuracy of a spindle in relation to a T-slot in the bed. A true arbor is inserted in the spindle and the contact point of the indicator bears against it. Any inaccuracy is shown on a greatly increased scale by the pointer, the end of which may be seen at the right end of the indicator body. While these two indicators differ in construction they operate on the same principle and are used for the same class of work. There are also many other forms or designs of this same general type.

Special Dial Indicating Gages. — The dial indicator is used in combination with many different gaging devices, for testing the accuracy of finished parts. Fig. 25 shows a gaging fixture which is used for testing the inside diameters of the inner races of ball bearings. The race to be tested is placed over a stud at the left end of the gage, as shown in the illustration. This stud has a two-point bearing and the gaging arm forms the third point. A multiplying lever extends to the other end of the fixture and

Fig. 25.   Internal Gaging Fixture for Ball Bearing Races

the end of this lever bears against the plunger of a dial gage, which shows any variation in the diameter. Errors above or below the standard size are multiplied ten times so that the gage, which normally reads to thousandths, gives a direct reading to 0.0001 inch. By adjusting the dial so that the hand points to zero, when the gage is set to the standard size, the amount of variation either above or below this standard dimension is easily determined. Thus it will be seen that gages of this type are " comparators " that show variations from a standard size but are not used for taking measurements.

Another form of dial gage for testing the outside diameters of finished ball bearings is shown in Fig. 26. This gage consists of a multiplying lever, one end of which comes into contact with the work while the other end bears against the plunger of the dial gage. The test is made by simply rolling the bearing on the true base of the fixture and under the end of the multiplying lever. Obviously, any variation from the standard size to which the gage is set, is indicated by the dial. The arm which carries the multiplying lever can be adjusted vertically in the slotted supporting bracket in order to set the gage for testing different sized bearings. The exact adjustment of the gage is obtained by comparing it with a master disk, such as the one shown to the right of the illustration. This disk is also used for checking the gage at intervals, to insure accurate readings.

Fig. 26. Gage for Testing Outside Diameters of Ball Bearings

A great many special gages of the same general type as those shown in Figs. 25 and 26 are now used, especially in inspection departments. A common idea of a gage is that it should have gaging surfaces which are a duplicate or exact complement of the part to be tested. A thread plug gage, for instance, is often regarded as being properly a steel plug threaded and hardened, the thread shape conforming exactly to that of the standard thread. While manufacturers furnish gages of this type in response to common demands, it is well known that such a gage is not a properly designed testing instrument. It is true that the ordinary thread plug gage may answer the purpose for which it was designed and it is also true that it is hardly practicable to devise a low-priced gage in which the faults of the plug gage are eliminated. The plug gage satisfies the common demand for a standard form that can be referred to for all dimensions, angles and shapes. A gage, however, which is used to test, at the same time, all the dimensions of even a simple part, is likely to be inaccurate and unreliable. As a general principle, a cylindrical plug gage should never be required to measure more than one diameter, and a solid gage should not be made to verify the concentricity of more than two cylindrical surfaces simultaneously. If a fixed gage is made to test several surfaces, it is impossible to determine definitely where the inaccuracies are; moreover, a gage of this type may seem to fit perfectly when in reality there are errors which remain undetected. A thread gage which is in the form of a threaded hole may seem to fit a screw perfectly and yet the screw may be several thousandths of an inch under size. For instance, if there is an error in the lead of the thread this may cause a screw that is under size to fit into the gage without, perceptible play.

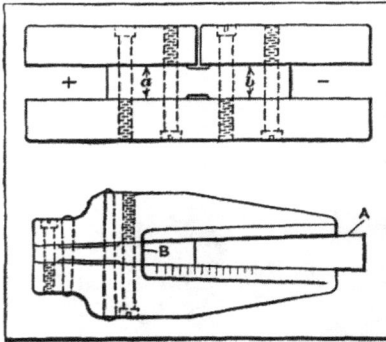

Fig. 27.   Sectional Gages

The type of gage having movable parts connecting with graduated dials, so that plus or minus readings can be taken directly, has replaced many gages of the fixed type, especially for inspection work, because they give a direct comparative measurement within very small limits of accuracy. Such gages, however, are often quite expensive and, in many cases, simpler forms serve all practical requirements.

Sectional Gages. — A sectional snap gage formed of four parts is shown in the upper part of Fig. 27. The measuring jaws, instead of being integral with the gage body, are attached to a central block by screws, as shown. The width a of one end of this central block equals the size of the " go " end of the gage; width b equals the size of the " not go " end. The gage jaws are made flat. The advantage of this design, as compared with a solid snap gage, is that when the accuracy is impaired as the result of wear the gage can be restored to its original accuracy by simply removing the gage jaws and truing them by grinding and lapping. The same principle can also be applied to an angular taper gage, as shown by the lower view, Fig. 27. The gage jaws are attached to a central block B finished accurately to the required taper, and the size of the work A is tested by pushing it between the jaws and noting the position of the end relative to a standard graduation mark. When the gage becomes inaccurate, as the result of wear, the jaws are removed and trued. A master plug should be used, occasionally, for testing the accuracy of the gage. By having one jaw graduated, as shown, the amount of inaccuracy may also be gaged, by noting how far the end of the work comes short of or projects beyond the standard dimension mark.

Precision Measuring Machines. — The measuring machine is an instrument of great precision that is used for originating standard lengths and for verifying the accuracy of reference gages. It might properly be defined as an instrument for obtaining accurate subdivisions of the standard unit which forms the basis of the system of measurement that is used. The Pratt & Whitney measuring machine is shown in Fig. 28. This machine has a heavy cast-iron bed upon which two heads are mounted. One of these heads A is normally fixed to the bed, whereas the other head B is adjustable along the accurately machined ways of the bed, for the measurement of various lengths. Each head has a spindle or measuring point and the part to be measured is supported between these spindles upon the rests C, which are of suitable shape at the top to center the work. Measurements up to 1 inch are obtained by means of a large graduated index wheel D, a scale and pointer at H being provided for approximate setting. For lengths greater than 1 inch, the sliding head is set by means of a standard bar E at the rear. (See Fig. 29.) The divisions or graduations, which are exactly 1 inch apart, are marked upon the surfaces of plugs set into this bar and are so fine that they are imperceptible to the naked eye. The sliding head is located for the inch positions by adjusting it with reference to these lines. In order to secure an adjustment which will exactly conform to the divisions on the standard bar, the sliding head is equipped with a powerful microscope F which is provided with a very fine line which is set with reference to the bar graduations. The screw of the sliding-head spindle, by means of which the adjustments for fractional parts

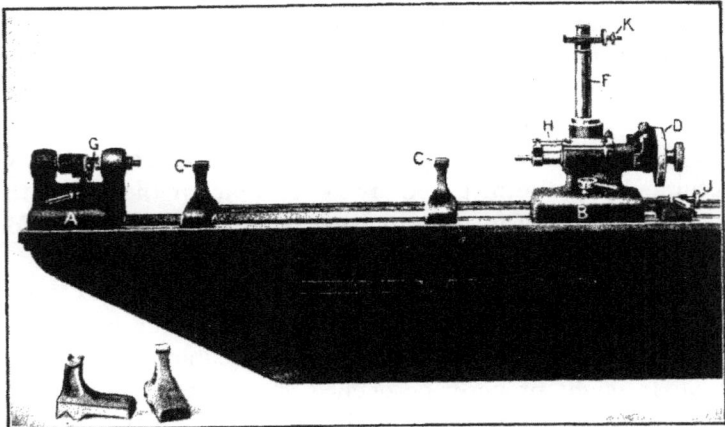

Fig. 28.   Precision Measuring Machine

of an inch are obtained, has twenty-five threads per inch, and the index wheel D has 400 graduations on a machine for English measurements; therefore, each graduation represents a $\frac{1}{400}$ of $\frac{1}{25}$ or 0.0001 inch, and the divisions can easily be subdivided into quarters or even less by estimation.

In order to insure a light contact or delicate and uniform pressure between the measuring points each time a measurement is taken, the machine is provided with a simple indicating device on the fixed head. This consists of two auxiliary jaws between which is held a small cylindrical plug G, by the pressure of a light helical spring, which operates the sliding spindle to which one of the jaws is attached. The tension of this spring is so adjusted that when the measuring points are not in contact the jaws will hold plug G in a horizontal position by friction. When the spindles are in perfect contact, either with each other or with the work, the tension on the spring is slightly reduced and plug G swings down to a vertical position, but any excess pressure will cause the plug to drop out of the jaws; hence, the contact for all measurements should be just enough to cause plug G to swing down to the vertical position.

How to Use a Measuring Machine. — To illustrate the application of the machine shown in Fig. 28, suppose it were necessary to set it for testing the accuracy of a special end-measuring bar

Fig. 29. Rear View of Precision Measuring Machine

or gage 10.2508 inches long. First the machine should be set in the zero position with the measuring points in contact. In order to do this, adjust the screw of the linear scale at the top of the head to zero, and set the pointer of the index wheel D nearly to zero; then slide the head until the measuring faces are almost in contact, and then by means of screw J, at the right of the head, adjust one spindle against the other until the indicating plug G shows a tendency to move from its horizontal position. Next clamp the head firmly and adjust the index wheel until the plug G swings down to a vertical position. Then set the adjustable index pointer to zero, and the line in the eye-piece of the microscope so that it exactly coincides with the zero line of the graduated reference bar E, Fig. 29, at the rear. The adjustment of the line in the eye-piece is made by means of screw K. The machine is now set in the zero position, and, when adjusting the head for the required measurement, care must be taken not to disturb the eye-piece of the microscope.

To measure from zero to one inch, the micrometer screw can be used direct, but for greater dimensions locate the sliding head so that the line in the eye-piece of the microscope coincides with the graduated plug from which the measurement is to be taken, the fine adjustment necessary being obtained by means of screw J at the right of the head. In this particular case, the head would be moved back along the bed until the line in the eye-piece of the microscope exactly coincided with the tenth graduation line. The distance between the measuring surfaces is now 10 inches. As the length required is 10.2508 inches, the screw would be turned back until the scale and index wheel of the adjustable spindle showed a movement of 0.2508 inch. As the pitch of the screw is $\frac{1}{25}$ inch, each complete turn of the index wheel equals 0.040 inch; hence, for a movement of 0.2508 inch, the turns of the index would equal 0.2508 ÷ 0.040 = 6.27, or six full turns and 108 divisions.

To test the rod, the index wheel would be turned a little beyond the required distance and the rod placed between the measuring surfaces. After setting plug G in a horizontal position, the index wheel would be turned back to the 10.2508 position. If plug G dropped before this position was reached, it would indicate that the rod was too long, but if it remained in a horizontal position, it would show that the rod was under size. In either case, the exact amount of error could easily be measured. When measuring an end gage, especially if of considerable length, care should be taken to prevent any variation in the temperature of the gage. When it is desired to test one gage with another master gage, the machine is first set by adjusting the contact points with the master gage. The other gage is then placed between the jaws and its length compared by referring to the graduations on the machine.

The Pratt & Whitney machines graduated for English measurements are standard at 62° F. It is not necessary, however, to use the machine at this initial temperature, because variations due to temperature changes will affect both the work and the machine practically the same, although when the machine is used for scientific research, the initial temperature should be adhered to.

Standard Unit of Length. — The international meter is the fundamental unit of length in the United States. The primary standard is deposited at the International Bureau of Weights and Measures near Paris, France. This is a platinum-iridium bar with three fine lines at each end; the distance between the middle lines of each end, when the bar is at a temperature of 0° C., is one meter by definition. Two copies of this bar are in the possession of the United States and are deposited at the Bureau of Standards, in Washington.

The United States **yard** is defined by the relation, 1 yard $= \frac{3600}{3937}$ meter. The legal equivalent of the meter for commercial purposes was fixed as 39.37 inches by law in July, 1866, and experience having shown that this value was exact within the error of observation, the United States office of standard weights and measures was, by executive order, in 1893, authorized to derive the yard from the meter by the use of this relation. No ultimate standard of reference for angular measurements is required, inasmuch as the degree can be originated by subdivision of the circle.

Tests at Bureau of Standards. — The Bureau of Standards employs various methods of making comparisons of bars which are submitted by manufacturers for test, the method depending upon the kind of bar, the accuracy desired, and the adaptability of the apparatus available to the bar or test piece. Thus, there are several classes of tests, such as Class A, for reference standards, Class B, for working standards, etc. The fee charged for this work depends, of course, upon the class and nature of the test. Metric length measures tested by the bureau are standardized at 20° C., and standards in the customary units of yards, feet, and inches are made to be correct at 62° F.

INDEX

www.ingramcontent.com/pod-product-compliance
Lightning Source LLC
Chambersburg PA
CBHW031953190326
41520CB00007B/232